Food Packaging and Safety Regulations

NIPA® GENX ELECTRONIC RESOURCES & SOLUTIONS P. LTD.
New Delhi-110 034

About the Editor

Dr. Tanweer Alam is presently working as an Additional Director & Regional Officer - Indian Institute of Packaging (IIP), Delhi, an autonomous Institute under Ministry of Commerce and Industry, Govt. of India. He is a member of Scientific Committee of Sustainable Packaging, FSSAI. He is also chairman & member of various Technical and Sectional Committees of BIS, CHD 16, CHD 15 & FAD 6. He was Rapporteur in FSSAI - Packaging Regulations. Dr. Alam is an Editior - in - Chief of Journal of Packaging Technology and Research (Springer Publication) and Associate Editor of Journal of Postharvest Technology and Asian Journal of Dairy and Food Research. He is also Editor of Packaging India, a Bimonthly Institutional Magazine and is in Editorial Board of Processed Food Industry. He has guided 5 Ph.D and a dozen of Post Graduate Students. He has been a Mentor for INSA & ICAR Fellows as well. Dr. Alam has published more than 100 research and technical papers in high impact referred journals, apart from delivering several invited lectures both in India and abroad. Spearhead for receiving 21 design patent, 1 granted Indian Patent and filed 3 Indian Patents. So far, he has published five books on Packaging and Edited dozens of Proceedings, Annual Reports, Vision Documents and Technical Souvenir. He was also the board member of World Packaging Organization (WPO) and Asia Packaging Federation (APF) and has worked as guest faculty at University of Copenhagen, Denmark under faculty exchange programme funded by European Union (EU). Dr. Alam has been a National Fellow of NADSI. He has been a jury member for Asia Star Award. Dr. Alam has organized more than 50 conferences / workshops of National / International stature (Asia Packaging Congress & Research Conclave, ISPI etc.) for promoting packaging and has received several awards and appreciations amongst which Exhibition and Business Excellence Award 2023 given by MSME Global Conference in the recent one. He is spearhead in taking initiative for commencing Six Months Online Certificate Programme on "Packaging & Processing of Fragrance, Flavour & Cosmetics. Dr. Alam is representing as a governing board member FFDC, Kannauj apart from APEDA & GS1. He is also the COURT member of prestigious Guru Gobind Singh Indraprastha University (GGSIPU). He has introduced BS and MS course in Packaging Technology and CPE course in IIP.

Food Packaging and Safety Regulations

Tanweer Alam
Additional Director & Regional Officer
Indian Institute of Packaging (IIP)
Delhi-110092

NIPA® GENX ELECTRONIC RESOURCES & SOLUTIONS P. LTD.
New Delhi-110 034

NIPA® GENX ELECTRONIC
RESOURCES & SOLUTIONS P. LTD.
101,103, Vikas Surya Plaza, CU Block
L.S.C. Market, Pitam Pura, New Delhi-110 034
Ph : +91-11-43860225, Mob.: +91 9717133558, 9540816132
E-mail: newindiapublishingagency@gmail.com
Website: www.nipaersources.com

Print ISBN: 978-93-58877-18-2
e-book ISBN: 978-93-58878-09-7

Composed and Designed by NIPA®.

Shri R.K. Mishra, IRS
Director
Indian Institute of Packaging

Foreword

Food Safety is a matter of Global concern that affects millions of people across the globe. To safeguard the society at large, Packaging laws & Regulations have been introduced. The present book "Food Packaging and Safety Regulations-An Update" by Prof. (Dr.) Tanweer Alam and other authors stands in good stead with the up-to-date information on various relevant topics related to food safety regulations.

Over the years, Indian Institute of Packaging (IIP) has played a pivotal role in inter aria enhancing the competitiveness through packaging improvements. The Institute's activities encompass the entire spectrum of packaging, including research and development, education, training, testing, information, consultancy and various other industry focused services.

I am sanguine that this book will be greatly beneficial for professionals related to food science technology and packaging technology.

R.K. Mishra, IRS

Editorial Board Member

Preface

Food Packaging has been used since ancient times to transport & store food. The safety of food packaging process needs to be paramount. Food Packaging & Safety Regulations book is focused on the various regulatory aspects related to food packaging. It covers wide range of packaging types and materials such as paper, metal, glass, plastic and there impact, hazards, safety concerns.

The book provides an insight about some recent updates to food packaging laws & regulations that may be useful and also provide an updated knowledge about the subject matter. It gives an overview of the food safety & standards, safety measures, Packaging designs, potential hazards & emerging trends in Food packaging.

This book is a culmination of years of expertise, insight and best practices gathered by professionals and scholars in the field of food packaging.

On behalf of the institute I extend my heartfelt gratitude to all those who have been part of the book.

I am confident the book shall provide the readers with sufficient information to understand the essential requirements.

Tanweer Alam

Contents

1

Regulatory Frameworks of Food Packaging in India

Vannie Dogra and Jasvir Singh

Global Regulatory Affairs, International Flavours & Fragrances, Gurgaon, India

Introduction

Food packaging is one of the most dynamic and fast-growing industries in the current scenario of increasing industrialization. With the constant upsurge in the food wastage at different stages of processing (suchas harvest, storage, processing, transportation, and distribution) leads to more than 15% of food produced ending up in the dumping grounds due to food spoilage caused by climatic conditions andmicroorganisms.Ever-increasing demands of consumers for convenient food makes packaging technology a need of the hour in order to keep the food safe, wholesome and of high quality at the same time.

So appropriate packaging has the potential of not only preventing spoilage but also taking care ofhealthand well-being ofthe individuals.Therefore, packaging plays the most important role in the overall safety of the food product.This makes packaging as the ever-evolving field since new requirements need new materials which demands new approaches and technology for better quality products.

By definition, Food packaging is a well-organized system of protecting the food product from any kind of physical, chemical, and biological damage in order to ensure that food reaches the consumer in its most safe condition at minimum applicable cost. It is bridge between production and consumption (Paine, 1992).

However from a regulatory perspective, Food Safety & Standards Authority of India defines "Food Package or container" as a pre-packaged box, bottle, casket, tin, barrel, case, pouch, receptacle, sack, bag, wrapper, or such other things in which an article of food is packed(https://www.fssai.gov.in/upload/uploadfiles/files/Gazette_Notification_Packaging_03_01_2019.pdf, 2019)

Levels of Packaging

There are different levels of packaging depending on the function it performs-

a) **Primary Packaging** means that the packaging material is in direct contact with the food product. Such material should not release any harmful substance in the food product during contact which is injurious to the health of the consumer (viz cheese in tin cans)

b) **Secondary Packaging** means packaging material which encloses series of primary food packages and does not come in direct contact with food products (viz cheese tins packed in corrugated paperboard boxes)

c) **Tertiary Packaging** means bulk packaging of the secondary packages generally termed as pelletizing (viz arranging the corrugated boxes through shrink wrapping film)

d) **Quaternary Packaging** means packing the pelletized loads into big metal containers for long distance transport by air, land, or sea.

Sometimes this categorization can vary depending the use of the packaging material. The same packaging material can also perform well on all the levels of packaging.

Functions of Food Packaging

The three major roles played by packaging are-

a) **Containment**- A package contains and holds the product facilitating its storage and distribution from manufacturing to the end consumer. Hence, the package needs to be of adequate shape and size, leak proof compatible with the product, grouped together to save space and cause no damage during transportation. It is a part of preservation process once the integrity of the package is breached, the product is no longer preserved.

b) **Protection**- Most vital part of packaging is to protect the product against any kind of spoilage, contamination, loss, and deterioration to maintain the quality and freshness. Three types of protection offered are-

- **Physical**- Safeguards the productsfrom fluctuating light, temperature,oxygen, and moisture conditions retaining the color, flavor, and nutrition value.
- **Biological**- Acts as a shield against insects, pests, rodents and microbial attacks.
- **Mechanical**- Prevents product from any kind of breakage, bruises, puncture, abrasion, spillage, bursting, crushing during handling, loading and transportation.

c) **Selling-** The way a product is designed, and information is presented on the label tells a lot about how the product is going to sell in the market.Product label acts as a salesperson. A well-informed label and an attractive package influences the consumer to buy the product. Also, the label should contain all the necessary information required for the consumer decision for comparison for ultimate buying and Instructions of use.

Package Designing

A well-designed package is vital for the success of the product. Essential factors required for creating an effective package are-

a) **Product Evaluation-** Considering the physical properties (such as nature of food, shape, size, texture, bulk density) and chemical properties (like sensitivity to light, temperature, oxygen availability, moisture gain or loss) in order to decide the right packaging material suitable for keeping the product intact and safe both internally & externally.

b) **Product Handling Hazards-** During loading, unloading, storage, transportation, and distribution a number of hazards come into picture which affects the overall quality of the food package. Therefore, a package must be constructed by keeping in view of all the foremost mechanical (deformation, vibration, compression, piercing), climatic (drip loss, condensation, high/low temperature) and biological hazards (bacteria, mold, fungi, insects, moths, rodents, termites, microorganisms) a product may face during its handling to prevent any sort of damage.

c) **Marketing needs-** Food Package is a silent communicator to the consumer. It not only brings reputation to the company but also acts as a promotional guide. A comparison with other products in market is inevitable, which drives the advertisement and marketing push, which a product will need in order to improve consumer awareness and market share. Hence, the kind of package, quantity, offers, mandatory labelling are certain things to be kept in mind to meet the sales objectives.

Therefore, the role of packaging and the surrounding where the package needs to perform its function should be focused for an effective package design.

Types of Packaging Materials

There are various types of packaging materials which are in use now a day. Most important ones are plastic polymers, paper, corrugated boxes, glass, tin, aluminum, tetra packs, retort pouches, laminates, shrink films, aseptic packaging, vacuum packaging and modified atmospheric packaging.

FSS (Packaging) Regulations, 2018 defines packaging material as materials such as cardboard, paper, glass, metal, plastic, multi-layer packaging material used for packaging of food products.(https://www.fssai.gov.in/upload/uploadfiles/files/Gazette_Notification_Packaging_03_01_2019.pdf, 2019)

Global Regulatory Frameworks of Food Packaging

Food packaging has been an easy way to transport and stock the goods from prehistoric periods. However, those were from natural sourceswhich are now shifted to need based packaging material such as plastic polymers, glass, metals, paper etc. With increase in the consumer awareness, the demands for longer shelf life and serving size-based food packages also rises which requires specific properties of packaging material. Every time a reason is added to this list, more strain is put on the new packaging forms.

Rapid increase in the use of packaging materials has also brought many other aspects to fore, such as problems of waste management, environmental impact, migration of substance, recyclability and risk of food contamination etc. These have also resulted in the need for a regulatory framework in these areas. Many countries have developed their own regulations while some have started following the rules and regulation of other global bodies. There is also a trend towards harmonization of regulations across countries, while some countries still stick to their own approaches.

Most countries have evaluated the following basic aspects in order to determine safety of packaging materials:

- type of materials
- leaching of substances
- import requirements
- Intent and objective of regulations

As regulations have to deal with multiple aspects of packaging materials, it is very difficult to incorporate all of them in just one regulatory framework. Hence in some scenario's food packaging is fully incorporated in the food law while in some others it is difficult to separate rules on food from rules on packaging. Some countries have just started to draft their packaging regulation whereas some others are updating their already existing regulations. One significant barrier for trans-border trade is that several nations issue their laws only in their national language, and official translations into a more accessible language such as English are not available (Veraart, 2010). This impact understanding and awareness of these laws among wider set of stakeholders.

Let's have a look at some major global frameworks.

i. **Codex Alimentarius Commission** CAC is an international food standards body established in May 1963 jointly by the FAO & WHO with the objective of protecting consumer's health and ensuring fair practices in food trade. Codex standards and guidelines are followed as reference standards by many countries as well as WTO. These standards are voluntary in nature, unless they are made mandatory by national governments. Codex have developed General Standard for the Labelling of Prepackaged Foods(Codex, revised in 2018) referred as CXS 1-1985 which lays down general and mandatory labelling guidelines for pre-packaged food products. However, it doesn't address the details of packaging materials.

ii. **European Union** European Union has established a self-governing risk assessment agency called "European Food Safety Authority" (EFSA), which has been tasked to undertake safety evaluation of all substances which are intended for food use. All the results from such evaluations are issued in the EFSA journal. These findings are then used for making regulations by European Commission, which acts as risk management agency for Europe. EU has issued a regulation dealing with packaging materials in contact with foods, and these cover general principles of safety and inertness for all food contact materials. Inertness means the food contact material should not release any constituents into food at levels which are harmful to human health and should not alter food composition, taste and odor in an unacceptable way (Legislation, n.d.). This framework lays down special requirements for active and intelligent materials and articles, requirements for authorization of substances, labelling and compliance related aspects (Regulation (EC) No 1935/2004 of the European Parliament and of the Council on materials and articles, 2004).

   Law for specific food contact materials have also been established in European Union, such as plastic materials, active and intelligent materials, recycled plastic materials, ceramics, and regenerated cellulosic film.(Commission Regulation (EU) No 10/2011 on on plastic materials and articles intended to come into contact with food, 2011)is the most comprehensive version which sets out rules for the composition of Food contact materials (FCMs), the list of substances permitted to be used in the making of plastic FCMs, limits on use of these materials and directions for compliance. This regulation is updated on regular basis and can be referred from (Legislation, n.d.).

Further, there are commission regulations on specific substances such as use of bisphenol A in varnishes and coating coming in contact with food and restricting use of epoxy derivatives in materials which tend to come in contact with food. In the absence of EU measures, Member state rules and regulations should be followed. A provisional list of food additives allowed to be used in plastic food contact materials is also available.

iii. **United States -Food and Drug Administration** FDA is responsible for protecting the public health by ensuring the safety of nations food supply. It is the major centralized agency which enforces laws and regulations for food and food additives. Packaging materials are covered under food contact substances which are defined by Food, Drugs and Cosmetics Act (FDCA) 1958 as any substance intended for use as a component of materials used in manufacturing, packing, packaging, transporting, or holding food if such use is not intended to have any technical effect in such food. Plastics, paper, coatings, adhesives are covered under food contact substances, which when mixed with other substances are called as food contact materials (FCMs). Code of Federal Regulations (C.F.R.) under Title 21 on food and drugs, Part 176-186 deals with FCMs, also referred to as indirect food additives which requires clearance for being considered as a food additive, as they are expected to migrate into the food. If they become part of the food, they are either authorized as a food additive as per the applicable food additive regulation, 21 CFR Part 170.3. If not this, then it is approved as an indirect food additive through 21 CFR 174-178. Since, above processes are time consuming, the substances are authorized by FDA through the submission of Food Contact notification by evaluating the safety of the material within 120 days or else issue a No Objection certificate. If a substance is "Generally Recognized as Safe" it may be approved by FDA through GRAS notification program. However, a GRAS substance can be exempted both from authorization and notification if the substance was a common food ingredient before 1958, falls under FDA listed GRAS substances and FDA approved GRAS Substances. Additionally, under Threshold of Regulation rule (TOR) non-carcinogenic substances with concentration below 0.5ppb are exempted from authorization. Under this rule, substances which earlier were direct food additive and when used in food contact materials will not result into exposure of more than 1% of ADI are also exempted. Thus, a substance can be considered as GRAS if not harmful to human health as per the intended end use and the FBOs need not require notifying FDA for use.

iv. **Food Safety Authority Australia and New Zealand (FSANZ)** Food contact substances are regulated under Australia New Zealand Food Standard Code 2000 which monitors functions of substances coming in contact of food. It lays down definition for food additive, processing aids and package. The code is quite general in nature, it does not specify permitted materials for use in food packaging. Rather the manufacturer should ensure that the packaging material is safe and follow all the applicable legislation. The code lists down certain requirements for packages to comply with such as any package or material in contact with food should not cause any harm to human health, not exceed the permitted levels of heavy metals and toxicants. In the guidance document, Safe Food Australia 2016 specifies that the packaging should comply with the maximum limits in food for food packaging chemicals and recycled & reused plastics can be used for food packaging, provided they won't contaminate food and are suitable as per the intended use. Regulatory framework for migration of chemical from packaging material in FSANZ to food is still being reviewed and updated.

v. **Japan** National Regulation and Industry voluntary rules are followed while regulating food contact materials in Japan. Food additives, packages, containers are regulated by Food Sanitation Act 1947. Japan independently don't have a positive list of substances for FCMs, but, industry voluntary rules contains a list of FCMs from Title 21 CFR, EU additive list etc.

Regulatory Framework of Food Packaging in India

Before 2006, there were various laws & regulation pertaining to food such as Prevention of food adulteration act, 1954 and food product orders for different food categories. Following these rules was very challenging for the FBOs since with they were supposed to follow multiple laws, report to multiple authorities and departments. To avoid this confusion, all the acts and orders were repealed into a single act referred as **Food Safety and Standards Act, 2006** which became a single reference point for food.

This led to the development of **Food Safety and Standards Authority of India (FSSAI)** to regulate the manufacture, storage, distribution, sale and import of food products as well as to ensure availability of safe and wholesome food for human consumption by setting globally benchmarked standards of food. FSSA Act, 2006 consist of 12 Chapters in which Chapter-IV covers General Provisions for Article of Food (Section 19-24). General requirements towards packaging and labelling of food is enclosed in Section 23 of the act.

Like all other sections, Section 23 of the act provides an overarching direction w.r.t. packaging materials. Clause 2 of section 23 clearly states thatevery single FBO shall ensure that the labelling and presentation of food which includes their shape, appearance or packaging, the packaging materials used, the manner in which they are arranged and the setting in which they are displayed, and the information which is made available about them through whatever medium, does not mislead consumers.

Initially in the PFA days, there used to be a combined rule dealing with both labelling as well as packaging. This approach was taken up as such by FSSAI, into **Food Safety and Standards (Packaging and Labelling) Regulation, 2011** which came into force on 5th August 2011. This regulation covered general packaging requirements for all products and specific packaging requirements for different food categories such as milk and milk products, fats & oils, canned products etc. This was followed by general and specific labelling requirements. However, this approach was not aligned to global approaches in this area.

Hence after a lot of discussions, FSSAI decided to develop a new approach, which was aligned to global practices. Hence,the combined packaging and labelling regulation was split into two separate regulations, each dealing with a specific area. This resulted in gazette notification of packaging regulation in 2018, followed by labelling and display regulation in 2020.

1. **Food Safety and Standards (Packaging) Regulations, 2018** (FSSAI,https://www.fssai.gov.in/upload/notifications/2019/01/5c2de53ee7a3bGazette_Notification_Packaging_03_01_2019.pdf, 2019)
2. **Food Safety and Standards (Labelling and Display) Regulations, 2020**(FSSAI, FSS (Labelling and Display) Regulations, 2020, 2020)

The basic structure of FSS (Packaging) Regulation, 2018 includes general &specific requirements forpackaging materials followed by Schedule I, II & III which list down Indian Standards to be followed when using paper& board, metal & alloys and plastic materials for food packaging. This is followed by schedule IV which suggests the type of packaging material as per the food product category.

FSS (Packaging) Regulation, 2018 specifies following general requirements for FBOs to ensure that the packaging material (PM) used shall be as per the packaging regulations.

General requirements of packaging material to be fulfilled by the FBOs are-

1. Every food business operator shall ensure that the packaging material used shall be in accordance with these regulations. Provided where Indian Standards are not available, then relevant International Standards may be complied with.
2. Any material in direct food contact used for packaging, storage, distribution of food shall be of food grade quality. Food Grade as defined in this regulation means material made of substances which are safe and suitable for their intended use and shall not endanger human health or result in unacceptable change in the composition of the food or organoleptic characteristics.
3. Packaging Material, sealing material and closure system should be compatible for the type of product and the outer conditions where a package performs its function.
4. It should be able to bear any kind of mechanical, chemical, temperature conditions during transportation.
5. Package should be clean, hygienic and tamper proof.
6. Re-using already used tin container not permitted.
7. Wherever plastic container (5*l* or above)& glass bottles are reused for packaging, ensuring the suitability and easy of cleaning is most important.
8. Printing inks for use in food packaging must comply to IS 15495 :2004, Code of Practice, Bureau of Indian Standards.(Standards, 2004)
9. Printed surface of food should not be in direct contact of food.
10. Newspaper or similar materials should not be used for wrapping of food since it leaches out the ink to food.
11. In case of Multilayer packages, the layer in direct contact with food shall meet the requirements laid down in Schedule I, II & III of this regulation.
12. Every FBO should obtain a certificate of conformity issued by NABL accredited laboratory for Packaging material/layers coming in direct food contact.

Regulation 4 refer to **Specific requirements for primary food packaging** intended to come in direct food contact-

1. **Paper & Board Materials-** Should be of uniform thickness, free of any visible specks, grease marks, cutes, pinholes & blemishes. Paper & board used for manufacturing of boxes, plates, cups, cartons, containers etc. should be of food grade quality, free from any kind of contaminants and conform to either of the Indian Standards mentioned in Schedule I.

Schedule – I

Paper and board materials intended to come in contact with food products

S. No	List of Standards
1.	Grease proof paper – IS 6622
2.	Vegetable parchment or Grease proof paper or Aluminum Foil Laminate – IS 7161
3.	Aluminum Foil Laminates for Packaging – IS 8970
4.	General purpose packing or wrapping Paper – IS 6615
5.	Folding Box Board, uncoated – IS 1776
6.	Corrugated Fibre Board Boxes- Specification (Part 1) – IS 2771

Note: The wax used for coating the paper or board shall be paraffin wax conforming to Type I of IS4654.(Gazette_Notification_Packaging_03_01_2019)

2. **Metal & Metal Alloys-** Containers which are rusty, chipped enameled container & copper, brass containers which are not properly tinned when used for packaging may render the food unfit for human consumption. Therefore, relevant grades of metal & alloys must be used for storing & packing of food products and must comply with either of the Indian Standards mentioned in Schedule II.(Gazette_Notification_Packaging_03_01_2019)

Schedule – II

Metal and Metal Alloys intended to come in contact with food products

S. No	List of Standards
1.	Cold-reduced Electrolytic Tinplate – IS 1993/ISO 11949
2.	Cold reduced Electrolytic Chromium or Chromium Oxide – Coated Steel - IS 12591/ ISO 11950
3.	Wrought Aluminum and Aluminum Alloy Sheet and Strip for General Engineering – IS 737
4.	Aluminum and Aluminum Alloy Bare Foil for Food Packaging – IS 15392
5.	Specification for Crown Closures – IS 1994
6.	Specification for Round Open Top Sanitary Cans for Foods and Drinks – IS 9396 (Part 1)
7.	Specification for Round Open Top Sanitary cans for Foods and Drinks – IS 9396 (Part 2)

3. **Glass Containers-** Should be free from any blisters, mold marks, stones, chipping, and other visible defects, have smooth surface without any cracks and sharp edges and the sealing surface should be free any hairline cracks and seam marks.
4. **Plastic Materials-** a) Plastic materials used for the manufacturing of food packages must comply with either of the Indian Standards mentioned in Schedule III provided-
   - Drinking water both packed & mineral water should be packed in transparent, colorless, tamper proof bottles and must conform to Indian Standards pertaining to the kind of packaging material used.

Type of Packaging Material	Indian Standard
Polyethylene (PE)	IS: 10146
Polyvinyl chloride (PVC)	IS: 10151
Polyalkylene terephthalate (PET and PBT)	IS: 12252
Polypropylene (PP)	IS: 10910
Food Grade Polycarbonate	IS: 14971
Polystyrene	IS: 10142

   - All pigments and colorants specified in Indian Standard IS:9833 are allowed in plastic container (5l or above) made of PET & Polycarbonate for packaging of drinking water.
   - Container transparency should not be less than 85% in light transmission.
   - Food grade packaging material which may or may not contain plastic as part of the packaging material but compatible with water can also be used. (FSSAI, https://www.fssai.gov.in/upload/notifications/2022/01/61f2431e10029Gazette_Notification_Water_27_01_2022.pdf)

b) Plastic originated packaging material shall pass through the overall migration limit of 60mg/kgor 10mg/dm2when tested as per IS 9845 with no visible color migration. They should not release the substances exceeding the specific migration limits mentioned in Table-1.(Gazette_Notification_Packaging_03_01_2019)

c) As per FSS (Packaging) Amendment Regulation, 2022Products made from recycled plastics which includes carry bags can be used for packaging, storing & carrying any articles of food products as and when standards are developed by the food authority. Such materials should comply with other applicable national standards/regulation. The same amendment implemented the approved guidelines for recycling of post-consumer PET acceptance criteria for recycled PET for food contact applications through Annexure-1(https://www.fssai.gov.in/upload/advisories/2022/01/61ea5c8e8713cDirection_Recycled_Plastics_19_01_2022.pdf)

In view of the above, a new direction have come which communicates that FBOs carrying out the recycling process should get the authorization from Food Authority. The mechanism of authorization is defined in the new order (FSSAI, https://www.fssai.gov.in/upload/advisories/ 2022/01/61eb9b40949b5Mechanism_Recycled_Plastics_22_01_2022.pdf).

Schedule – III

Plastic Materials intended to come in contact with food products

S.No.	List of Indian Standards
1	Specification for Polyethylene for its safe use in contact with foodstuffs, pharmaceuticals and drinking water – IS 10146
2	Specification for Polystyrene for its safe use in contact with foodstuffs, pharmaceuticals and drinking water –IS 10142
3	Specification for Polyvinyl Chloride (PVC) and its copolymers for its safe use in contact with foodstuffs,pharmaceuticals and drinking water – IS 10151
4	Specification for Polypropylene and its copolymers for its safe use in contact with foodstuffs, pharmaceuticalsand drinking water – IS 10910
5	Specification for Ionomer Resins for its safe use in contact with foodstuffs, pharmaceuticals and drinking water– IS 11434
6	Specification for Ethylene Acrylic Acid (EAA) copolymers for their safe use in contact with foodstuffs, pharmaceuticals and drinking water – IS 11704
7	Specification for Polyalkylene Terephthalates (PET & PBT) for their safe use in contact with foodstuffs,pharmaceuticals and drinking water - IS 12252
8	Specification for Nylon 6 Polymer for its safe use in contact with foodstuffs, pharmaceuticals and drinking water– IS 12247
9	Specification for Ethylene Vinyl Acetate (EVA) copolymers for its safe use in contact with foodstuffs,pharmaceuticals and drinking water – IS 13601
10	Specification for Ethylene Metha Acrylic Acid (EMAA) copolymers and terpolymers for its safe use in contactwith foodstuffs, pharmaceuticals and drinking water – IS 13576
11	Specification for Polycarbonate Resins for its safe use in contact with foodstuffs, pharmaceuticals and drinkingwater – IS 14971
12	Specification for Flexible Packaging Materials for packaging of Edible Oils, Ghee and Vanaspati - IS 14636
13	Specification for Polyalkylene Terephthalates (PET & PBT) for Molding and Extrusion – IS 13193

14	Specification for Polyethylene Films and Sheets – IS 2508
15	Specification for Linear Low-Density Polyethylene (LLDPE) Films – IS 14500
16	Specification for High Density Polyethylene Materials for Molding and Extrusion – IS 7328
17	Specification for Melamine-Formaldehyde Resins for its safe use in contact with foodstuffs,pharmaceuticals and drinking water – IS 14999
18	Low Density Polyethylene Films – IS 2508
19	Blow Molded Polyolefin Containers - Part 2 : Over 5 Liters, up to and Including 60 Liters Capacity – IS7408
20	Stretch Cling Films – IS 14995

Regulation 5 pertains **to suggestive list of packaging material for food product categories provided in Schedule IV**. However, this list does not restrict to use any other packaging material compatible for the product other than mentioned in Schedule IV by complying with the relevant standards. (Gazette_Notification_Packaging_03_01_2019)

Indian Standards referred in FSS (Packaging) Regulation, 2018 are developed by **Bureau of Indian Standards** which is a National Standards Body of India established **under BIS Act 2016** for the harmonious development of the activities of standardization, marking and quality certification of goods and for matters connected therewith or incidental thereto.(BIS, The Bureau of Indian Standards Act , 2016). BIS frames Indian Standards in alignment with the national priorities for various sectors in which Food and Agriculture falls as one of the 14 other departments. There is a list of Indian Standards provided on the BIS website which specifies department wise Standards which can be referred.(BIS, https://www.bis.gov.in/wp-content/uploads/2021/07/listofproducts.pdf).

Apart from FSSAI and BIS, **Legal Metrology (Packaged Commodities) Rules, 2011 under Legal Metrology Act, 2009** which lays down labelling requirements of the food packages meant for retail sale and wholesale packages. Rule 5 represents Specific commodities to be packed and sold in recommended standard packages for sales and distribution given in Schedule II. This schedule provides commodities to be packed in specified quantities. Commodities covered under this includes Baby food, biscuits, breads, butter margarine, cereals, pulses, coffee, tea, powdered beverages, edible oils, salt mineral water, drinking water, aerated soft drinks etc.

Where a commodity is packed in sizes other than what is Specified in Schedule II shall bear a statement on the label that'Not a standard pack size underthe Legal Metrology (Packaged Commodities) Rules, 2011 or 'nonstandard size underthe Legal Metrology (Packaged Commodities) Rules,2011'.

Apart from the above regulations which deals with packaging materials as well as labelling, environmental aspects of packaging materials have also become an important area of concern for governments. In line with global trends, Government of India has also issued Plastic Waste (Management and Handling) Rules, 2011 under Ministry of Environment, Forest, and Climate Change. These provide regulatory framework for plastic waste generated and plays a vital role for sustainable environment through waste management. Plastic Waste Management (Amendment) Rules, 2018 lay down conditions of manufacture, sale, distribution & use of carry bags, plastic sheets, multi layered packaging-

1. Material and bags should either be in natural shade i.e. without any added pigments or should contains pigments & colorants allowed in IS 9833:1981titled as "List of pigments and colorantsfor use in plastics in contact with foodstuffs, pharmaceuticals and drinking water".
2. Carry bags made of recycled plastics should not be less than 75 microns in thickness effective from 30 Sep' 2021 and 120 microns in thickness effective from 31Dec'21.
3. Plastic sheets not becoming important part of multi-layered packaging and plastic sheet covers used for packing , wrapping should not be less than 50 microns in thickness except where such thickness could impair the product functionality.
4. Manufacturer should not provide plastic as a raw material to the producer if don't have a valid registration from state pollution control boards.
5. Sachets using plastic materials should not be used for packing, selling, and storing tobacco products.
6. Carry bags and commodities made from compostable plastic should conform to Indian Standard IS 17088:2008 - Specifications for Compostable Plastics. Manufacturer and seller should obtain a certificate from Central Pollution Control Board for carrying forward this activity,
7. Plastic waste recycling should follow Indian Standard: IS 14534:1998 named as "Guidelines for Recycling of Plastics", as revised from time to time.

As we can see regulatory frameworks for packaging materials have continued to evolve day-by-day, and harmonization of regulations from global standards perspective has been gaining significance in industry. We do expect that this will continue to result in an improved safety of the end products as well as a regulated business practices.

Future Regulatory Developments

1. We do expect more and more harmonized approaches to develop across the world, including in India.
2. Environmental aspects like waste management and sustainability concerns to gain even more significance in coming days.
3. Safety evaluation principles to evolve further so that they can address new scientific developments around packaging materials, such as active packaging materials. Currently, safety evaluation is built around the principle of testing neutrality and inertness of packaging materials. Lower the general and specific migration levels, safer is the packaging material. However, with the advent of active packaging materials, this principle will need to evolve further, to address active packaging materials which are actually designed to release active components into food to provide some additional benefits.
4. Use of new packaging materials such as nanomaterials will also require some adjustments to safety evaluation processes and protocols.

References

Act, F.S. (n.d.). https://www.japaneselawtranslation.go.jp/law/detailmain?id=12&vm=2&re#:~:text=Food %20 Sanitation %20Act & text=Article %201% 20 The % 20 purpose % 20 of, to %20 protect %20 citizens' %20 good %20 health.

Authority, E. f. (n.d.). Retrieved from https://www.efsa.europa.eu/en.

3. BIS. (2016). The Bureau of Indian Standards Act. MINISTRY OF LAW AND JUSTICE, 1-17.

BIS. (n.d.). https://www.bis.gov.in/wp-content/uploads/2021/07/listofproducts.pdf.

Board, C. P. (n.d.). https://cpcb.nic.in/rules-4/.

Board, C. P. (n.d.). https://cpcb.nic.in/uploads/plasticwaste/Notification-12-08-2021.pdf.

Brennan, J. G. (2010). Food Processing Handbook, ISBN: 978-3-527-32468-2 (Second Edition ed.).

CFR. (n.d.). 170.39 Threshold of regulation for substances used in food-contact articles, https://www.ecfr.gov/current/title-21/section-170.39.

CFR. (n.d.). PART 184 - Direct Food Substances Affirmed As Generally Recognized As Safe, https://www.ecfr.gov/current/title-21/part-184.

CFR. (n.d.). PART-182- Substances Generally Recognized As Safe, https://www.ecfr.gov/current/title-21/part-182.

Code of Federal Regulations (C.F.R.) under Title 21 on food and drugs, Part 170.3. (n.d.).

Code of Federal Regulations (C.F.R.) under Title 21 on food and drugs, Part 176-186 . (n.d.).

Codex. (revised in 2018). https://www.fao.org/fao-who-codexalimentarius/sh-proxy/en/?lnk=1&url=https%253A%252F%252Fworkspace.fao.org%252Fsites%252Fcodex%252FStandards%252FCXS%2B1-1985%252FCXS_001e.pdf.

Comission, E. (n.d.). Retrieved from https://ec.europa.eu/food/safety/chemical-safety/food-contact-materials_en.

Commission Regulation (EU) 2020/1245. (2 Sep' 2020). Amending and correcting Regulation (EU) No 10/2011 on plastic materials and articles intended to come into contact with food, https://eur-lex.europa.eu/legal-content/EN/TXT/?uri=CELEX:32020R1245. Official Journal of the European Union.

commission regulation (EU) No 10/2011 on on plastic materials and articles intended to come into contact with food. (2011, January 14). Official Journal of the European Union.

FDA. (n.d.). Generally Recognized as Safe (GRAS), https://www.fda.gov/food/food-ingredients-packaging/generally-recognized-safe-gras.

FDA, U. (n.d.). Inventory of Effective Food Contact Substance (FCS) Notifications, https://www.cfsanappsexternal.fda.gov/scripts/fdcc/index.cfm?set=fcn.

FDCA, x409(h)(6), 21 U.S.C. x348(h)(6). (n.d.).

FSANZ. (n.d.). Food Standards Code, https://www.foodstandards.gov.au/code/Pages/default.aspx.

FSSAI.(2019).https://www.fssai.gov.in/upload/notifications/2019/01/5c2de53ee7a3bGazette_Notification_Packaging_03_01_2019.pdf.

FSSAI. (2019, January 03). https://www.fssai.gov.in/upload/uploadfiles/files/Gazette_Notification_Packaging_03_01_2019.pdf.

FSSAI. (2020, November 17). FSS (Labelling and Display) Regulations, 2020. Gazette Notifications, 48. 24. FSSAI.(n.d. .https://www.fssai. gov.in/upload/advisories/2022/01/61ea5c8e8713cDirection_Recycled_Plastics_19_01_2022.pdf.

FSSAI.(n.d.).https://www.fssai.gov.in/upload/advisories/2022/01/61eb9b40949b5 Mechanism_Recycled_Plastics_22_01_2022.pdf.

FSSAI.(n.d.).https://www.fssai.gov.in/upload/notifications/2022/01/61f2431e10029 Gazette_Notification_Water_27_01_2022.pdf.

Gazette_Notification_Packaging_03_01_2019. (n.d.).

https://ec.europa.eu/food/system/files/2016-10/cs_fcm_legis_additives-prov-list.pdf. (n.d.).

https://eur-lex.europa.eu/legal-content/EN/TXT/?uri=CELEX:32018R0213. (n.d.).

Jelen, P. (1985). Food packaging technology. In: Introduction to Food Processing. Reston.

Legislation, E. (n.d.). https://ec.europa.eu/food/safety/chemical-safety/food-contact-materials/legislation_en.

Ministry of Environment,F. a.-I.-3.-s. (1-20). Plastic Waste Management (Amendment) Rules, 2018, https://cpcb.nic.in/displaypdf.php?id=cGxhc3RpY3dhc3RlL1BXTV9HYXpldHRlLnBkZg==.

Paine, F. A. (1992). A Handbook of Food Packaging .

Regulation (EC) No 1935/2004 OF THE EUROPEAN PARLIAMENT AND OF THE COUNCIL on materials and articles. (2004, October 27).

Robertson, G. (1992). Food Packaging — Principles and Practice. Marcel Dekker. New york.

Sina Ebnesajjad, F. G. (2016). Regulatory Aspects of Food Packaging- A global Matter. In E. b. Imprint: William Andrew, Multilayer Flexible Packaging, ISBN-978-0-323-37100-1 (pp. 1-43). doi:https://doi.org/10.1016/C2014-0-02918-8

Standards, B. o. (2004). https://law.resource.org/pub/in/bis/S02/is.15495.2004.pdf. IS 15495 (2004): Printing Ink for food packaging - Code of practice, 10. Retrieved from IS 15495 (2004): Printing Ink for food packaging - Code of practice.

The Legal Metrology (Packaged Commodities) Rules, 2011,http://wbconsumers.gov.in/writereaddata/ACT%20&%20RULES/Act%20&%20Rules/9%20The%20Legal%20Metrology%20(Package%20Commodities)%20Rules,%202011.pdf. (2009). The Legal Metrology Act, 1-43.

Veraart, E. b.-R. (2010). Global Legislation for Food Packaging Materials. WILEY-VCH Verlag GmbH & Co. KGaA,.

2

Chemical Migration and Food Packaging

Farheen Khan, V.P. Sharma and Tanweer Alam[#]

CSIR-Indian Institute of Toxicology Research, Vishvigyan Bhawan 31 M.G. Marg, Lucknow, Uttar Pradesh

[#]Director, Indian Institute of Packaging [IIP], Andheri Mumbai, Maharashtra

Abstract

In the fast-moving consumer goods (FMCG) industry, polymeric packaging offers a multitude of benefits. While safe and elegant, its improper disposal poses a significant threat. Plastics are incredibly useful and versatile due to their light weight, strength, aesthetics, and variety in colors, sizes, shapes, and prices. Innovative active packaging solutions with added features, such as extended shelf life or antimicrobial properties, are used to attract consumers in the perfume, cosmetics, food, and beverage industries.

However, some packaging materials can leach harmful chemicals, including heavy metals, exceeding safe limits and potentially causing health problems. The potential for consumer exposure to harmful chemicals needs thorough evaluation using OECD guidelines to ensure safety for all age groups. Exposure to certain chemicals in packaging materials can lead to health problems such as reproductive disorders, diabetes, neurodegenerative diseases, kidney problems, and allergies. The continued use of plastic food packaging poses a health risk to humans.

Stakeholders involved in packaging need proper training to handle, package, and transport goods according to regional regulations and international standards. This will enhance quality control and ensure compliance at the manufacturing level. The packaging sector needs systemic innovation to promote sustainability by addressing plastic pollution and chemical migration. Following food management system guidelines will enhance understanding, leading to economic growth and improved quality of life.

Keywords: *Biomagnifications, Endocrine Disruptors, Heavy metals, Packaging, Plastics & Polymers*

Introduction

Packaging of food material is an essential component of food industry since packaging provides protection, physical tampering resistance and unique appearance to the food products which attracts the customers (Robertson, 2006). The handling of food products would be inefficient and costly if they are not packed properly (Robertson, 2006). The style and design of packaging influence their sale as well as demand. The main aim of packaging is to corroborate the safety and quality of food by preventing microbial contamination (Lee K.M, 2010). Plastics, papers, glass and metals are the common materials used for packaging of food products (Arvanitoyannis and Bosnea, 2004). Among these plastics are most commonly used packaging material because of their low cost, light weight, design freedom and high durability (British Plastic Federation). Various additives are added to the plastics to achieve desired physical and chemical properties. These additives are stabilizers, plasticisers, lubricants and flame retardants. These additives have potential to transfer from the packaging materials to the food stuffs(K Bunia *et al.*, 2013).

The efficient food packaging solutions are the need of the hour for the Earth. The food losses including the production, post harvest, processing stages of the supply chain and food wastage includes loss of edible food occurring at the final stages of food chain that is during retail supply or ultimate consumption. The food born illnesses cannot be ignored which are a consequence of the contamination of food by bacteria, viruses, parasites, new chemical entities or toxicants. We need to optimise the food supply chain, increase food shelf life and consciousness amidst food producers to food consumers. The active and intelligent packaging may help the globe through sufficient food for the increasing population. Scavengers/ absorbers are important solutions which may absorb substances from the packaging inner atmosphere (ethylene absorbers, oxygen scavengers; emitters and antioxidants).

Nowadays, advance communication system is feasible through smart packaging for communicating about the product quality, brand consumer connection, tracking, theft and counter fitting conducts in the food supply chain (Schyns, Z.O. *et al.*, 2021). The characteristics like relative humidity, temperature, microorganism's growth may help to assure the quality of the packaged product and estimated shelf life. The indicators/sensors may help us to provide visual qualitative or semi quantitative relevant information through differential colour intensities or irreversible changes.

The material development and over arching strategies for sustainability is critical option. Due to paradigm shift in biobased items or paper-based packaging solutions is the increasing trend.The intelligent packaging is useful in monitoring the quality of the food product. Intelligent packaging consists of three components data carriers, indicators and sensors. The environmental conditions are monitored through sensors. Temperature, humidity and acidity of the food product can be analysed through the sensors. Indicators ensure the quality of the packed food and data carriers store and transfer data (Müller, P. *et al.*, 2019). Conventional packaging leads to the migration of chemicals.

The migration of monomers (styrene, vinyl chloride), direct and indirect additives (plasticisers, stabilisers, heavy metals) may lead to toxic implications. It alters the food quality in terms of aroma and flavour of the food products. Chemicals like Bisphenol A (BPA) and phthalates have been found to migrate from the packaging materials into the foodstuffs. BPA is a known endocrine disruptor which affects the reproductive system of animals as well as humans (Vilarinho, F *et al.*, 2019). Certain heavy metal like cadmium, lead, and chromium also migrates from the packaging materials. The heavy metals when present in excess cause neurological disorders such as Alzheimer's disease (Restriction of the Use of Certain Hazardous Substances in Electrical and Electronic Equipment, 2011). Therefore, various agencies such as US EPA (Environment Protection Agency) have limited the use of multilayer packaging for foodstuffs.

It is essential to ensure the safety of food additives, enzymes, flavouring agents, processing aids, and ingredientsintentionally added to food items. Besides this safety of food packing and other food contact materials becomes vital as the processes of food preparation, processing, and preservation have multiple critical points. The migration test is conducted in food production facilities so as to determine whether chemical substances are being transferred from food packaging and production equipment to avariety of packaging solutions.

Migration of Chemicals from Packaging Materials

The migration /leaching of the chemical moieties are dependent on the simulants, duration, surface area of contact and temperature under specific conditions of exposure. Thus, migration is dependent on the type o polymeric materials used or packaging and the pH of the stored in food items viz namely juices, water, milk and other eatables.

In recent years, trace metals in packaged foods have caused widespread concern about their health implications. Research have shown that trace metals can cause impaired fertility, increased spontaneous abortion, neurological and immune system disorders and cardiovascular ailments (Yüzbasi *et al.*, 2003).

Mechanism of Migration

Diffusion is the main mechanism by which the migration of mass movement of molecules occurs between the packaging material and the foodstuffs. Chemicals move from the area of their highest concentration to the region of their lowest concentration until an equilibrium is reached (Arvanitoyannis, I.S *et al.*, 2014).

The rate of molecular diffusion is calculated by Fick's second law:

$dCp/dt = D\ (d^2\ Cp/dx^2)$

Where Cp: concentration (mg/g) of migrant in packaging material

D: coefficient of diffusion (cm^2/s)

t(s): time of diffusion

x: distance (cm) between food and packaging material (Silva *et al.*, 2007).

Monitoring the factors which influence migration would control contamination of food and food safety.

Types of Migration

Migration based onthe number of migrants

Migration comprises of two terms, overall migration and specific migration Overall migration includes the sum of all the migrating substances from unit area of packaging materials whereas specific migration involves the migration of single chemical moiety. The quality of packaged foodis evaluated by considering both types of migration (Robertson, 2006). The specific migration limits (Max mg/Kg) are as per FSSAI 2018 regulations for barium (01.00), cobalt (00.05), copper (05.00), iron (48.00), lithium (00.60), manganese (00.60) and zinc (25.00) respectively.

Migration on the basis of nature of food packed

Migration is divided into 3 categories depending on the nature of food-

Non migrating system –in this system pigments or inorganic substances are migrated in minute quantities and the migration of high molecular weight packaging polymer is more.

Volatile migrating system – here the migration from dried products occurs where there is no contact of the food with the packaging. The flux of volatile substances occurs by diffusion or evaporation, desorption from and adsorption to products.

Leaching migrating system –in this system the food must be in direct contact with the packaging material for the process of leaching. The transfer is initiated by the process of diffusion, then dissolution finally dispersing into the food product. Leaching of materialsfrom plastic packaging to semisolid or fluid food items corresponds to leaching migration system (Lee *et al.*, 2008).

Gas phase migration

This type of migration requires the gaseous medium for exchange and involves the permeation of substances from the printed side of the packaging (Karen *et al.*, 2006).

Set-off migration

This type of migration involves the transfer of inks and varnishes used for printing. Initially they migrate from the outer layer which is printed to the inner layer and then from the inner layer to the food products through direct contact. This type of migration can be visible depending on the substance used for printing (Karen *et al.*, 2006).

Condensation or distillation migration

The shelf stability of the food is enhanced by heating or sterilisation. But certain substances migrate during the process of boiling or sterilisation of packed foods or food stored in cartons. Specifically, the volatile component migrates through the process of distillation from the steam released by the aqueous food (Karen *et al.*, 2006).

Factors Affecting the Process of Migration

Since migration of the substance is a complex phenomenon, several factors may affect the process of migration. The factors include nature of food, type, temperature and duration of contact, nature of packaging material and migrant characteristics. Food composition and nature are vital factors in evaluating migration of substances. Foods with high fat content intensify the migration process. Several studies were conducted to investigate the extent of migration of food stimulants. Based on these studies' different authorities in Europe and U.S have recommended several food stimulants (Triantafyllou *et al.*, 2007). Numerous studies have suggested that the extent of migration is related to direct or indirect contact. Direct contact enhances the migration rate whereas indirect contact involves the medium of gas for migration and therefore migrates slowly (Anderson and Castle, 2003). Migration of substances also depends on the time during which the food was in contactwith the packaging. Experimentally the migration of substance is proportional to the square root of the duration of contact (Arvanitoyannis and Bosnea, 2004). Temperature at which the food is stored influence the rate of migration. Generally, a higher temperature increases the migration rate until the equilibrium is achieved (Triantafyllou *et al.*, 2007). Migration of specific additives is significantly affected by the thickness and plasticisation of the packaging material. Thicker packaging slows down the rate of migration as compared to thinner layer of packaging (Nerin *et al.*, 2007). Microstructure of a substance also influence mass transfer rate for example spherical or branched molecules or molecules with or without side chains. Branched molecules have low migration rate as compared to unbranched molecules (Triantafyllou *et al.*, 2005).

Types of Migratory Compounds from the Food Packaging Materials

From printing inks

The single layer packaging material is extensively printed to show case the description of the product to the consumer. Eatables stored in this type of packaging are at a higher risk of contamination as the probability of migration of printing dyes increases which deteriorates the quality of food. In packaging generally,the printing is done through ultraviolet (UV)-curable inks. The printing of most of the single layer plastics is appropriately chosen so that its constituents do not migrate from the upper exposed surface to the inner stored in material. However, there are potential risk of migration. is exposed to a UV source which converts the photo initiator into free radical. These free radicals react with the monomers and initiate the process of polymerisation. During polymerisation these polymers interact with the pigments producing a good quality print on the packaging material. (Samonsek and Puype, 2013). Defective functioning of UV source increases the formation of free radicles which migrate to the packed food and compromise the safety of the consumers. Migration of benzophenone is also reported in several studies which is used as an odourless photo initiator in printing (Bradley *et al.*, 2013).

From adhesives

Sealing in multilayer packaging is done through adhesives. There are various types of additives available in the market such as polyurethane that solidifies through chemical reaction, acrylic and vinyl adhesives which harden by water (Aznar, M. *et al.*, 2011). Poly urethane is widely used for flexible laminates as it provides chemical and temperature resistance and thus serves as excellent adhesive. Adhesive manufacturers conducted a survey and listed more than 360 substances that have potential to migrate from the adhesive to the food. Use of laminates along with adhesive significantly increases the risk of health problems (Hoppe *et al.*, 2016).

Plastic packaging

Plasticizers

They are added to the polymer to increase the flexibility of the product. The most common plasticizer used in PVC are phthalates. There are various phthalates that are approved by regulatory agencies like dibutyl phthalate (DBP), diisononyl phthalate (DINP) and butyl benzyl phthalate (BBP). Presence of phthalates in the urine of several patience led to the confirmation of their migration from the packaging material as food was the major factor for the route of exposure of phthalate (Carlos, K.S *et al.*, 2021).Since phthalates are low molecular weight substances, they can easily migrate from the packaging to the food. (Pedersen *et al.*, 2008; De Toni *et al.*, 2017). In a study the assessment of migration regulatory of phthalates has been studied in plastic lunch boxes. The DBP and DIBP has been determined in the concentration of 1.5 mg/Kg and 2.4 mg/Kg respectively. The first order kinetic migration model was established with reference to phthalates (Liu, J.M *et al.*, 2020).

Thermal stabilizers

Thermal stabilizers help to prevent thermal degradation of the plastic. Epoxidized seed and vegetable oils (esterified soyabean oil) are widely in food packaging plastics. The use of stabilizers is not recommended particularly in food packaging because of toxic nature (K. Bhunia *et al.*, 2013).

Slip additives

Oleamides, stearyl erucamide, stearamide, erucamide and oleyl palmitamide are used as slip additives in the plastic packaging. They act as lubricants and prevent sticking between the layers (O.W. Lau *et al.*, 2000). The additive migration from low density polyethene were examined and found to be 88% for oleamide, 98% for erucamide, and 95% for stearamide respectively (Arvanitoyannis and Bosnea, 2004).

Light stabilizers

They are added to increase the longevity of the product for long term use. Hindered amines are used to prevent photo oxidation of the polymers (Poulos, A., 2021). These amines can be quantified by ultra-performance liquid chromatography and can be used as a tool to monitor compliance with regulatory limit. Benzophenone and benzotriazoles are other commercially available light stabilizers (Noguerol-Cal *et al.*, 2010).

Antioxidants

A plastic polymer oxidises in the presence of UV light and gets degraded. Use of antioxidant reduces the degree of oxidation. There are many commercially available antioxidants that are used in food packaging like Tinuvin P and Irganox168(Nestmann *et al.*, 2005). Certain vitamins can also be used as antioxidants like vitamins A, C and E and their derivatives like carotenoids. Metal ions like selenium are essential for the proper functioning of enzymes which participate in antioxidation reactions.Several studies have reported that synthetic antioxidants like butylated hydroxytolueneand propyl gallatemigrate significantly from the packaging material into the food stuffs (Papas, 2012).

Solvents

Solvents are used to enhance the dispersion of printing inks, that are used in plastic packaging. The solvents are prepared by mixing several organic compounds such as alcohol. Ketone and ester. (Boon, 2008). While printing, the residues of these solvent get entrapped in the packaging material and migrates when the food is being packed. The rate of migration of solvents largely depends on the concentration and its distribution. Solvent's migration ultimately changes the organoleptic properties of food beside being harmful to the environment (Robertson, 2006).

Monomers and oligomers

Monomers and oligomers act as a building block and undergo polymerisation reaction to produce different kinds of polymers. Styrene is the monomer employed to produce polystyrene and therefore used in packaging materials and when the food items come in contact with packaging material, they migrate into the food stuffs. Polystyrene is the common ingredient of packaging materials for the huge number of dairy products like ice cream,yogurt, and fruit juices (Tawfik and Huyghebaert, 1998). Studies have reported that breakdown of styrene monomer led to the formation of respective oxides which are potential mutagens and when metabolised in body, can lead to the formation of hippuric acid. Exposure of styrene may cause irritation of skin, eyes and affect the proper functioning of the central nervous system (CNS) (Tang *et al.*, 2000; Sadighara, P *et al.*, 2022).

Isocyanates

Isocyanates are used in the formation of polyurethanes. They are added to additives for the manufacturing of packaging materials. Aromatic primary amines are considered as subcategories of this class of compounds. Chemical leaching is observed in different materials like rubber, azo dye and polyurethane. The maximum concentration of isocyanates used in packaging material is < 1.0 mg kg^{-1} (Lau and Wong, 2000).

Vinyl chloride

Vinyl chloride is a colourless gas under normal pressure and temperature conditions but under the conditions of high pressure it is compressed into liquid form. The liquid form is used to manufacture polyvinylchloride (PVC) packaging materials (Robertson, 2006). It is highly toxic and therefore its use should be limited. The Agency for Toxic Substances and Disease Registry (2006), in U.S. have reported the daily exposure of vinyl chloride isless than 0.0004μg kg^{-1}.

Acrylonitrile

Acrylonitrile is used to manufacture ABS (acrylonitrile–butadiene–styrene) copolymer. It provides heat and chemical resistance to the ABS polymer. ABS polymer is used for the packaging of butter and therefore they were shown to leach from the ABS polymer (Yam, K.L *et al.*, 2012).

Polyethylene terephthalates

Polyethylene terephthalates (PET) is used to manufacture bottles to store fresh juice, milk, carbonated beverages and water. It is thermo polyester formed by the condensation of ethylene (Kim and Lee, 2012). PET is flexible and therefore it is quickly molded into desired shape. It is temperature resistant because of which the containers are used for heating of food. Antimony trioxide is a classified carcinogen, which is used as a catalyst in the production of PET. They are shown to migrate into the food stuffs with the detection limit of 0.06% and 1% (depending on the concentration of PET) in various beverages T (Nerín *et al.*, 2013; Silano *et al.*, 2008; Tsochatzis, E.D *et al.*, 2022).

Implications of Metals migrating from packaging

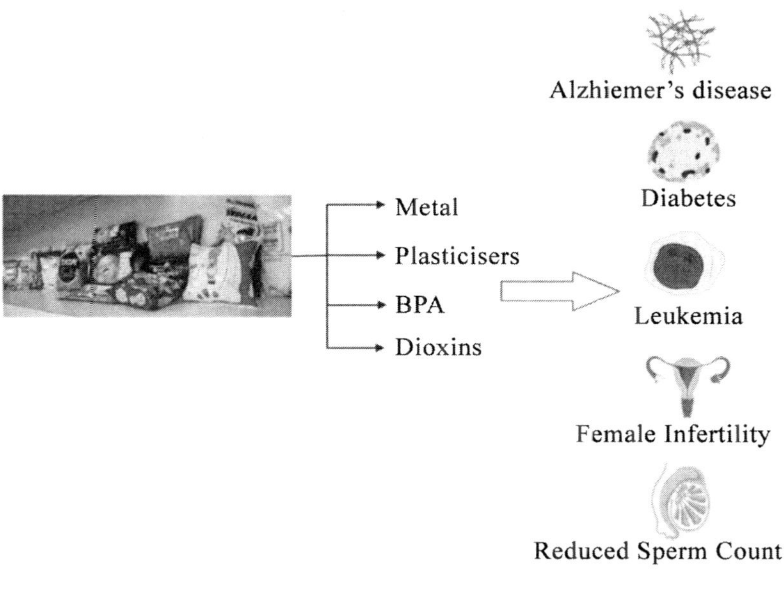

Figure 1: Migrants from Packaging Material and Implications

Tin (Sn)

Cans manufactured from tin are used to store food items and various carbonated drinks. Thin layer tin protects the metal can from corrosion. Traces of tin are shown to migrate into the food stuffs from metal packaging. Foods that contain high amount of tin (approximately 500mg kg^{-1}) may lead to gastrointestinal abnormality (Omori *et al.*, 1973; Benoy *et al.*, 1997). Studies have shown that threshold for an acute level toxicity occurs after above the dose of 730 mg kg^{-1}. Lacquer coating is preferable because it minimises the rate of migration of tin (Oldring, 2007).

Lead (Pb)

Lead is extensively used in the beverage containers. Lead causes organ toxicity particularly central nervous system (CNS). The rate of retention of lead is higher in Infant's brain and bones therefore they are at a greater risk of lead toxicity. Even low concentration of lead can cause mental retardation and diseases like Alzheimer's in adults (Skrzydlewska *et al.*, 2003; Robertson, 2006; KHAN, S *et al.*, 2022).

Aluminium (Al)

Aluminium is used to prepare multilayer food packaging which is also known as laminate. It is used along with several metals like copper, zinc, silicon, magnesium, manganese and iron in the form of alloy to design the packaging material (Taylor, 1964). Since aluminium is not essentially required for the functioning of the body, high intake and elevated levels of aluminium in tissues can cause disorders like dialysis encephalopathy, osteodystrophy, and microcytic anaemia. Migration of aluminium from metal packaging is influenced by the acidity of the food packed and solubility of the salt formed (metal packaging for food stuffs, 2007).

Chromium (Cr)

Apart from lead chromium migrates from ceramic and glass vessels into the food simulants (Szynal, T*et al.*, 2016). Coating of chromium by electroplating is done in tin-based cans to increase the stability against oxidative damage. It is highly toxic and causes undesirable sensory properties. Hexavalent chromium is hemotoxic, genotoxic and carcinogenic. It leads to oxidative stress that damages kidney and liver (Dayan *et al.*, 2000).

Copper (Cu)

Copper nanoparticles are used in the food packaging to enhance the food safety and shelf life as they show antimicrobial activity. Despite several advantages they are toxic to humans as copper nanoparticles migrate from coated nanofilms into the food stuffs (Hannon, J.C *et al.*, 2016).

Zinc (Zn)

Zinc oxide nanoparticles are used in the food packaging for several advantages such as blocking ultraviolet radiations and antimicrobial activity. Zinc is found to leach from the nanocomposite, particularly inlow density polyethene (LDPE) into the food simulants like distilled water (Bumbudsanpharoke, N *et al.* ,2019).

Paper packaging

Dioxins

Dioxins form a class of chemicals which include synthetic polychlorinated compounds such aspolychlorinated dibenzofurans and dibenzo-dioxins.They are by-products formed during smelting and chlorine bleaching of paper pulp. They are highly mutagenic and isomer called 2,3,7,8-tetrachlorodibenzo-para-dioxin is the most hazardous dioxin known (Ackermann *et al.*, 2006).

Benzophenone

Benzophenone acts as photo initiator in the processing of inks and lacquers. They are also used as wetting agent to improve the flowability of dyes and pigments (Anderson and Castle, 2003).Migrations of benzophenone takes place from outer side to the inner side of the cardboard at the time of stacking and therefore potentially contaminate the food products. 4-methoxybenzophenone is reported as carcinogenic and mutagenic. The migration of 2-hydroxybenzophenone is particularly observed for the food simulant Tenax (Elizalde, M.P *et al.*, 2020).

Nitrosamines

Nitrosamines are found incardboard and papers. These chemicals include morpholine which have the potential to migrate at the time of storage and processing involved in packaging. N-nitrosodimethylamine (NDMA) is highly toxic. It is a known carcinogen and induces tumours in various organs like liver, lung, kidney, pancreas and bladder (Mahan Hazarika, 2020).

Glass containers

Usually, glass is resistant to hydrophilic solutions and organic substances but when the it is autoclaved, the migration of silica takes place. The rates of migration are lower in case of glass containers and have minor influence on the organoleptic properties of the food (Guadagnino, E *et al.*, 2022). Since glass is inert material, the chemical leaching occurs mainly from inner surface to the foodstuffs. However, when the oily foods are packed in the glass container, they get contaminated by the migration of plasticizer from the closure (Food packaging forum, 2022).

Legislations for the Food Packaging

Materials that are used in food packaging must be safe for the consumer's health and must not alter the basic constituents of stored in food materials. Moreover, it should comply with the updated regulations of the Food Safety Standards Authority of India (FSSAI) and other associated guidelines issued by international organizations viz food and agriculture organization (FAO), European Union (EU), European food safety authority (EFSA), International Standardization Organisation (ISO), Environmental Protection Agency (EPA), Codex Alimentarius and World Health Organization (WHO). In order to prevent any potential transfer to foods,it is advisable to follow the principles of ISO 22000, hazard analysis and critical control point (HACCP), and good manufacturing practices (GMP).

The regulations of substances used in food contact articles (FCS) need to be holistically and internationally finalised for the harmonization of data acceptance. The resolutions related to them need special emphasis on following aspects

- Migration limit of BPA lowering to 0.05 mg/Kg.
- Ban on use of bisphenol A

Across Europe, packaging is governed by Commission Regulation No 10/2011 for plastic materials that are intended to come in direct contact with the food (Wu, Y., 2014; EU Commission Regulation, 2020. For the substances that are not included in the list, the migration should be less than 0.01mg/Kg (ratio 6:1) for food or food simulants. The directive 94/62/EC include all the packaging materials available in the market and the generated waste. It recommends the use of biodegradable plastic and encourages to prevent the use of toxic substances (Ebert, A. *et al*., 2017)

The German packaging act is incurred to prevent the negative impact of the packaging on the environment. To implement the act various regulations were laid such as curtailment of the use of lead, mercury, and cadmium in packaging materials andlabelling of the packaging material. The single use plastics are promoted so as to reduce the impact of the plastic. The oxo-degradable plastics are also banned as they break down into the micro-particles and are difficult to dispose (Heike Blank *et al*., 2021).

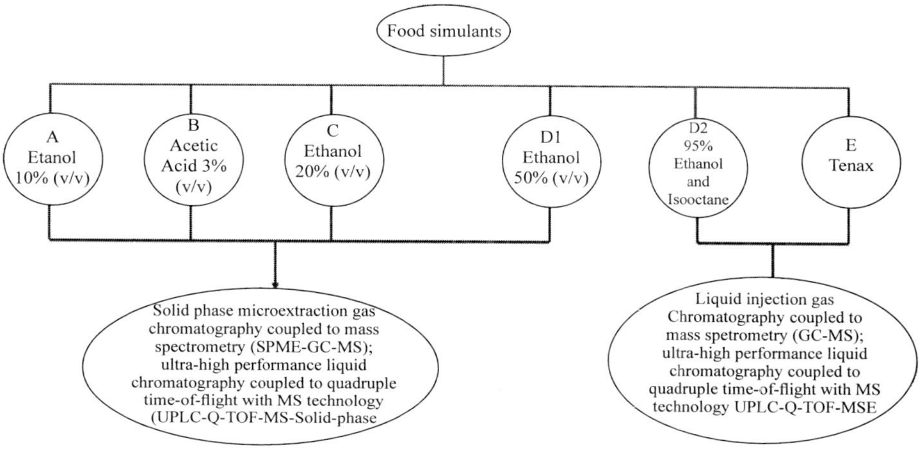

Figure 2 : Analytical methods used for different food simulants

Regulations for food packaging in India

Food Safety and Standards Authority of India (FSSAI) govern the packaging of different food items. The aims of the organisation include ensuring the safety of food, laying down standards based on scientific knowledge for the food related articles and regulating the manufacturing, distribution and sale of food.

Containers manufactured from plastic should comply with the Indian Standards Specification as enlisted below

S.No	Indian Standards	Title/ Details of Specification
1.	IS 10146:1982	Specification for polyethylene used in food packaging
2.	IS 10142:1999	Specification for styrene polymers used in food packaging
3.	IS 10151:1982	Specification for polyvinyl chloride used in food packaging
4.	IS 10910:1984	Specification for polypropylene used in packaging
5.	IS 11434:1985	Specification for ionomer resins used in food packaging
6.	IS 11704:1986	Specification for ethylene acrylic acid copolymer
7.	IS 12252:1987	Specification for poly alkylene terephthalates (PET)
8.	IS 12247:1988	Specification for nylon 6 polymer
9.	IS 13601:1993	Specification for ethylene vinyl acetate (EVA)
10.	IS 13576:1992	Specification for ethylene metha acrylic acid (EMAA)

Standards for paper and board material intended to be used in packaging are listed below

S.No	Indian standards	Title/ Details of Specification
1.	IS 6622: R 1999	Standard for grease proof paper
2.	IS 7161 :2013	Standard for vegetable parchment
3.	IS 8970: 1991	Standard for aluminium foil laminate
4.	IS 6615 :R1996	Standard for general purpose packaging or wrapping paper
5.	IS 2771:1990	Standard for corrugated fibre board boxes

Standards for metal and metal alloys intended to be used in packaging

S. No	Indian standards	Title/ Details of Specification
1.	IS 1993: 2006	Standard for cold-reduced electrolytic tin plate
2.	IS 12591:2006	Standard for cold reduced electrolytic chromium or chromium oxide coated steel
3.	IS 15392:2003	Standard for aluminium and aluminium alloy bare foil for food packaging
4.	IS 1994:1987	Specification for crown closures
5.	IS 9396:2012	Specification for round open top sanitary cans for foods and drinks

Analytical Approaches

Food quality and packaging are closely interconnected and are integral part of our daily life. The advancement and creativity in packaging solutions is making the assessment of migrants a challenge for the researchers. However new hyphenated and state of art analytical methods with better reproducibility and sensitivity are being used for the detection of possible food toxicants from the packaging origin. The leachatemay be due to printing inks, polymer degradation products or aromatic volatile compounds. These may be responsible for change in organoleptic properties of food.The potential leachate may include intentionally and non-intentionally added substances, monomers, nano particles, oligomers, antimicrobials and antioxidants. The selection of a particular analytical method may be influenced by the type of analysing samples, analytes and speciation. The transfer of leachate into food is always not similar for different type of polymeric materials. Thus, the spectrum of possible leachateis quite wide and few of them may be extra ordinarily toxic. In few cases of packaging materials there is significant variation in set-off migration from the outer part to the inner part and thus specific migration of chemical entities from packaging is of great relevance.

Volatile Compounds

The hybrid techniques like GC-Q-Orbitrap-MS are used for qualitative and quantitative analysis of volatile organic compounds (VOCs) in migration assay samples. These analytical techniques are highly sensitive and are capable of analysing analytes even in minute quantities (Lucera, A *et al.*, 2016;Cruz, R.M.S *et al.*, 2019). The electron impact ionization (EI) method is commonly used in GC–MS, where analytes are ionized by electrons with the energy of 70 eV. As this technique causes fragmentation of molecules, it is generally not recommended. The type of mass spectrometry analyser determines the resolution power of GC-MS. High-resolution mass spectroscopy (HRMS) is used to determine the molecular weight of the analytes in case of orbitrap. On the other hand, low resolution mass spectrometers (LRMS) determine the molecular weight in amu (atomic mass unit). The usage of these instruments is high because of their low cost, easy and cheaper maintenance.APGC is more sensitive as compared to the mass spectroscopy as the fragmentation of analytes is low. The degree of fragmentation depends on the energy of ionizing electrons. Here, Sensitivity is inversely proportional to fragmentation (Su, Q.-Z.*et al.*, 2019).

Another powerful technique is automatic multiple dynamic hollow fibre liquid-phase micro extraction (HFLPME), used to determine the migrants from active materials that contain essential oil. It is also used to analyse heavy metals and drugs by coupling with other analytical instruments. It is extensively used as it is cheap, sensitive and versatile (Oliveira, É.C *et al.*, 2014).

Non-volatile Compounds

Liquid chromatography coupled to mass spectroscopy (LC-MS) is extensively used for the analysis of non-volatile compounds even at very low level. It provides good resolution of peaks, particularly for the screening of nontarget compounds from migrating samples (Tigrine, A *et al.*,2019). To measure the accurate mass of the analyte, high resolution mass spectrometry techniques are used. This technique enables us to determine the elemental structure of the analyte.

Nanoparticles

Nanotechnology has revolutionized the food packaging by increasing the shelf life of the packed food. Nanoparticles are added to the polymer matrix to improve the barrier properties of the packaging materials. Nanoparticles can also migrate from the packaging material into the food stuffs and therefore are analysed through sensitive analytical techniques likemass spectrometry (ICP-MS) and electron microscopy (SEM and TEM) (Poças, F *et al.*, 2018). ICP-MS is frequently used for the analysis of nanoparticles along with its size and elemental composition. Moreover, the imaging of the nanoparticle can be done through SEM and TEM. These techniques have poor sensitivity at low concentration of nanoparticles (De la Calle *et al.*, 2018).

Toxicity of the Migrating Chemicals from Plastic Packaging

The chemicals that leach out from the packaging materials are collectively referred as "chemicals of concern" as they bio accumulate and are toxic to human health (Hahladakis, J.N. *et al.*, 2018). These chemicals have hazardous properties like being carcinogenic, mutagenic, reprotoxic or endocrine disruptors. Endocrine disruptors interfere with the normal functioning of hormones. Some known endocrine disruptors are Bisphenol A (BPA), phthalates,dioxins etc that migrate from the plastic into the packed food. Certain plasticizers and food additives are associated with diabetes. They alter the adipogenesis, normal functions of pancreatic β cell and hypothalamic neuropeptide which disrupt metabolic balance(GBD *et al.*, 2015).

Endocrine disruptors are also reprotoxic as they can either bind to estrogen receptors or increase aromatase activity and in turn increase estrogen sensitivity, or indirectly by their effect on gonadotropin releasing hormone leading to elevated levels of endogenous estrogen (Morgan M *et al.*, 2016). They affect the menstrual cycle, alter fertility and oogenesis and are correlated with diseases such as polycystic ovarian syndrome (PCOS) and endometriosis (Hauser R *et al.*, 2016). Apart being reprotoxic they are also related to neurodegenerative disorders like anxiety and depression, Parkinson's disease and multiple sclerosis (Rebolledo-Solleiro, D *et al.*, 2021). Apart from endocrine disruption activity certain chemicals like dioxins and its variant 2, 3,7,8-tetrachlorodibenzo-p-dioxin (TCDD) are carcinogenic in nature. They show leukemogenicity and increases the risk of acute myeloid leukaemia (AML) in TCDD exposed individual (Shallis, R.M *et al.*, 2022).

Phthalates have a very rapid metabolism in human body as they have a half-life of only 12 h. First, they are hydrolysed in the phase 1 metabolism and then they are conjugated with glucuronic acid in phase 2 metabolism to form glucuronide conjugate. Ultimately, they are released into the urine. The low molecular weight phthalates like DEP have an adverse effect on skin, nasal cavities and conjunctiva. Whereas high molecular weight phthalates cause methylation in genes that participate in androgen and estrogen response. Phthalates are associated with type 2 diabetes, obesity and asthma. The reprotoxic problems include early menopause, abortion and preterm birth (Wang *et al.*, 2021).

There is increasing concern about potential confounding from few chemicals and misclassifications due to variability in product (synthetic perfumes and regulated active ingredients). The intentionally added substances are not tested for their toxic nature and the exposure data are mainly based on assumptions. Thus, there are unpredictable outcomes for the chemical leachate from the packaging (Muncke, J *et al.*, 2020).

Conclusion

Despite a growing emphasis on recycling and circular economies, the global demand for plastic is projected to rise. By 2030, estimates suggest plastic circulation could reach 417 million tonnes annually. A major concern is the migration of unintended substances from plastic food packaging into food, with studies finding them in various materials like polyurethane additives and recycled materials. Further research is needed to pinpoint how these substances, like phthalates, move through the food chain.

Safety assessments using advanced techniques can help identify potential health risks. Interactions between packaging materials, food contents, and additives can also lead to unforeseen issues. Additionally, regulations from the EU and other governing bodies complicate matters by making it difficult to determine the chemical structures of contaminants in food packaging.

Socioeconomic factors and stakeholder cooperation play a role in adopting new packaging technologies. Ultimately, achieving sustainable packaging solutions requires a multifaceted approach. This includes focusing on renewable, recyclable, and reusable materials that minimize waste and environmental impact, while prioritizing product quality, food safety, and reducing food loss.

References

Ackermann, P.W., Herrmann, T., Stehr, C., Ball, M., 2006. Status of the PCDD and PCDF contamination of commercial milk caused by milk cartons. Chemosphere. 63, 670–675.

Agency for Toxic Substances and Disease Registry (ATSDR), 2006. Toxicological profile for Vinyl Chloride. U.S. Department of Health and Human Services, Public Health Service, Atlanta, GA.

Anderson, W.A.C., Castle, L., 2003. Benzophenone in carton-board packaging materials and the factors that influence its migration into food. Food Additives and Contaminants. 20 (6), 607–618.

Arvanitoyannis, I.S. and Kotsanopoulos, K.V., 2014. Migration phenomenon in food packaging. Food–package interactions, mechanisms, types of migrants, testing and relative legislation—a review. Food and Bioprocess Technology, 7(1), pp.21-36.

Arvanitoyannis, I.S., Kotsanopoulos, K.V., 2014. Migration Phenomenon in Food Packaging. Food-Package Interactions, Mechanisms, Types of Migrants, Testing and Relative Legislation–A Review. Food Bioproc. Tech. 7, 21–36.

Arvanitoyannis and Bosnea L,2004. Migration of substances from food packaging materials to foods. Crit. Rev. Food Sci. Nutr.

Aznar, M., Vera, P., Canellas, E., Nerín, C., Mercea, P. and Störmer, A., 2011. Composition of the adhesives used in food packaging multilayer materials and migration studies from packaging to food. Journal of Materials Chemistry, 21(12), pp.4358-4370.

Benoy, C.L., Hooper, P.A., Sneider, R., 1997. The toxicity of tin in canned fruit juices or solid foods. Food Chem. Toxicol. 9, 645–656.

Bradley, E.L., Stratton, J.S., Leak, J., Lister, L., Castle, L., 2013. b. Printing ink compounds in foods: UK survey results. Food Addit. Contam. 6 (2), 73–83.

British plastic federation.https://www.bpf.co.uk/

Bumbudsanpharoke, N., Choi, J., Park, H.J. and Ko, S., 2019. Zinc migration and its effect on the functionality of a low density polyethylene-ZnO nanocomposite film. Food Packaging and Shelf Life, 20, p.100301.

Carlos, K.S., de Jager, L.S. and Begley, T.H., 2021. Determination of phthalate concentrations in paper-based fast food packaging available on the US market. Food Additives & Contaminants: Part A, 38(3), pp.501-512.

Committee, E.S. Scientific Opinion on Exploring options for providing advice about possible human health risks based on the concept of Threshold of Toxicological Concern (TTC). EFSA J. 2012, 10, 2750.

Cruz, R.M.S.; Rico, B.P.M.; Vieira, M.C. 9-Food packaging and migration. In Food Quality and Shelf Life; Galanakis, C.M., Ed.; Academic Press: San Diego, CA, USA, 2019; pp. 281–301. ISBN 978-0-12-817190-5.

Dayan, A. D.; Paine, A. J. (2001). "Mechanisms of chromium toxicity, carcinogenicity and allergenicity: Review of the literature from 1985 to 2000". Human & Experimental Toxicology. 20 (9): 439451. Doi:10.1191/ 096032701682693062. PMID 11776406. S2CID 31351037.

De la Calle, I.; Menta, M.; Klein, M.; Séby, F. Study of the presence of micro- and nanoparticles in drinks and foods by multiple analytical techniques. Food Chem. 2018, 266, 133–145.

De Toni, L., F. Tisato, R. Seraglia, M. Roverso, V. Gandin, C. Marzano, R. Padrini, and C. Foresta. 2017. Phthalates and heavy metals as endocrine disruptors in food: A study on pre-packed coffee products. Toxicology Reports 4:234–9. doi: 10.1016/j. toxrep.2017.05.004.

Ebert, A.; Franz, R.; Gehring, C.; Kemmer, D.; Welle, F. Testing migration from food packaging materials. In Food Packaging Materials; CRC Press: Boca Raton, FL, USA, 2017; pp. 251–302.

Elizalde, M.P., Aparicio, J.L. and Rincón, M., 2020. Interpretation of the migration of benzophenone type photoinitiators into different food simulants and foodstuffs in terms of the physicochemical properties of the migrants. Food Packaging and Shelf Life, 23, p.100444.

EU Commission Regulation (EU) No 10/2011 of 14 January 2011 on Plastic Materials and Articles Intended to Come into Contact with Food. Available online: https://eur-lex. europa.eu/eli/reg/2011/10/oj (accessed on 8 February 2020).

European Parliament, Directive 2011/65/EC of the European Parliament and of the Council of 8 8 June 2011 on the Restriction of the Use of Certain Hazardous Substances in Electrical and Electronic Equipment, 2011, Available at: http://eur-lex.europa.eu/legal-content/en/ TXT/? uri=CELEX%3A32011L0065, L174 (2011) 88-110.

GBD 2013 DALYs and HALE Collaborators, Murray CJ, Barber RM, Foreman KJ, Abbasoglu Ozgoren A, *et al.* Global, regional, and national disability-adjusted life years (DALYs) for 306 diseases and injuries and healthy life expectancy (HALE) for 188 countries, 1990-2013, quantifying the epidemiological transition. Lancet. (2015).

Guadagnino, E., Guglielmi, M. and Nicoletti, F., 2022. Glass: The best material for pharmaceutical packaging. International Journal of Applied Glass Science.

Hahladakis, J.N., Velis, C.A., Weber, R., Iacovidou, E. and Purnell, P., 2018. An overview of chemical additives present in plastics: Migration, release, fate and environmental impact during their use, disposal and recycling. Journal of hazardous materials, 344, pp.179-199.

Hannon, J.C., Kerry, J.P., Cruz-Romero, M., Azlin-Hasim, S., Morris, M. and Cummins, E., 2016. Human exposure assessment of silver and copper migrating from an antimicrobial nanocoated packaging material into an acidic food simulant. Food and Chemical Toxicology, 95, pp.128-136.

Hauser R, Gaskins AJ, Souter I, Smith KW, Dodge LE, Ehrlich S, *et al.* Urinary phthalate metabolite concentrations and reproductive outcomes among women undergoing in vitro fertilization, results from the EARTH study. Environ Health Perspect. (2016).

Heike Blank, Nikolas Gregor. Plastic and Packaging Laws in Germany. 2021.

Hoppe, M., de Voogt, P., Franz, R., 2016. Identification and quantification of oligomers as potential migrants in plastics food-contact materials with a focus in polycondensates–A review. Trends Food Sci. Technol. 50, 118–130.

Jane Muncke. Migration. Food packaging forum. 2022.

K. Bhunia, S.S. Sablani, J. Tang, B. Rasco, Migration of chemical compounds from packaging polymers during microwave, conventional heat treatment, and storage, Compr. Rev. Food Sci. Food Saf. 12 (2013) 523–545.

Karen, A., Barnes, C., Sinclair, R., 2006. Chemical migration and food-contact materials. CRC Press, Boca Raton, FL.

Khan, S. and Khan, A.R., 2022. Migrating Levels of Toxic Heavy Metals in Locally Made Food Packaging Containers. Egyptian Journal of Chemistry, 65(1), pp.1-2.

Kim, D.J., Lee, K.T., 2012. Determination of monomers and oligomers in polyethylene terephthalate trays and bottles for food use by using high performance liquid chromatography electrospray ionization–mass spectrometry. Polym. Test. 31, 490–499.

Lau, O., Wong, S., 2000. Contamination in food from packaging material. J. Chromatogr. A. 882, 255–270.

Lee, D., Yam, K., Piergiovanni, L., 2008. Food Packaging Science and Technology. CRC Press, Boca Raton.

Lee K.M.2010.quality and safety aspects of meet production as effected by various physical manipulations of packaging materials.

Liu, J.M., Li, C.Y., Yang, F.E., Zhao, N., Lv, S.W., Liu, J.C., Chen, L.J., He, Z., Zhang, Y. and Wang, S., 2020. Assessment of migration regularity of phthalates from food packaging materials. Food science & nutrition, 8(10), pp.5738-5747.

Lucera, A.; Conte, A.; Del Nobile, M.A. Chapter 25—Volatile compounds usage in active packaging systems. In Antimicrobial Food Packaging; Barros-Velázquez, J., Ed.; Academic Press: San Diego, CA, USA, 2016; pp. 319–327. ISBN 978-0-12-800723-5.

Mahan Hazarika, 2020. Nitrosamine – A big no for printing inks for food and pharma packaging. India's premier magazine for modern packaging.

Metal packaging for food stuffs. International Life Sciences Institute (ILSI).2007.

Morgan M, Deoraj A, Felty Q, Yoo C, Roy D. Association between exposure to estrogenic endocrine disruptors-polychlorinated biphenyls, phthalates, and bisphenol A and gynecologic cancers-cervical, ovarian, uterine cancers. J. Carcinogen. Mutagen. (2016).

Muncke, J., Andersson, AM., Backhaus, T. *et al.* Impacts of food contact chemicals on human health: a consensus statement. Environ Health 19, 25 (2020).

Müller, P. and Schmid, M., 2019. Intelligent packaging in the food sector: A brief overview. Foods, 8(1), p.16.

Nerin, C., Contin, E., Asensio, E., 2007. Kinetic migration studies using Poropak as solid-food stimulant to assess the safety of paper and board as food packaging materials. Anal. Bioanal. Chem. 387, 2283–2288.

Nerín, C., Alfaro, P., Aznar, M., Domeño, C., 2013. The challenge of identifying nonintentionally added substances from food packaging materials: A review. Anal. Chim. Acta 775, 14–24.

Nestmann, E.R., Lynch, B.S., Musa-Veloso, K., Goodfellow, G.H., Cheng, E., Haighton, L.A., Lee-Brotherton, V.M., 2005. Safety assessment and riskbenefit analysis of the use of azodicarbonamide in baby food jar closure technology: putting trace levels of semicarbazide exposure into perspective e a review. Food Addit. Contamin. 22 (9), 875–891.

NICNAS, 2000. Acrylonitrile. Priority Existing Chemical Assessment Report. National Industrial Chemical Notification and Assessment Scheme (NICNAS). No.10.

Noguerol-Cal, R., López-Vilariño, J.M., Fernández-Martínez, G., González-Rodríguez, M.V., Barral-Losada, L.F., 2010. Liquid chromatographic methods to analyze hindered amines light stabilizers (HALS) levels to improve safety in polyolefins. J. Sep. Sci. 33, 2698–2706.

O.W. Lau, S.K. Wong, Contamination in food from packaging material, J. Chromatogr. A 882 (2000) 255–270.

Oldring, P.K.T., 2007. Exposure–the missing element for assessing the safety of migrants from food packaging materials. In: Barnes, K.A., Sinclair, R., Watson, D. (Eds.), Chemical migration and food contact materials. Woodhead Publishing, Cambridge (UK), pp. 122–157.

Oliveira, É.C.; Echegoyen, Y.; Cruz, S.A.; Nerin, C. Comparison between solid phase microextraction (SPME) and hollow fiber liquid phase microextraction (HFLPME) for determination of extractables from post-consumer recycled PET into food simulants. Talanta 2014, 127, 59–67.

Omori, Y., Takanaka, A., Tanaka, S., Ikeda, Y., 1973. Experimental studies on toxicity of tin in canned orange juice. J. Food Hyg. Soc. Jpn. 14 (1), 69–74.

Papas, A.M., 2012. Antioxidants in plastic packaged materials 1999 Online. Eastman Chemical Company.

Pedersen, G.A., Jensen, L.K., Fankhauser, A., Biedermann, S., Petersen, J.H., Fabech, B., 2008. Migration of epoxidized soybean oil (ESBO) and phthalates from twist closures into food and enforcement of the overall migration limit. Food Addit. Contam. 25 (4), 503–510.

Poulos, A., 2021. The Plastics Revolution. In The Secret Life of Chemicals (pp. 33-48). Springer, Cham.

Poças, F.; Franz, R. Chapter 10-Overview on European Regulatory Issues, Legislation, and EFSA Evaluations of Nanomaterials. In Nanomaterials for Food Packaging: Materials, Processing Technologies, and Safety Issues; Cerqueira, M.Â.P.R., Lagaron, J.M., Pastrana Castro, L.M., de Oliveira Soares Vicente, A.A.M., Eds.; Elsevier: Cambridge, UK, 2018; pp. 277–300. ISBN 978-0-323-51271-8.

Re-evaluation of Some Organic Chemicals, Hydrazine and Hydrogen Peroxide IARI Monographs, Volume 71(1999).

Rebolledo-Solleiro, D., Flores, L.C. and Solleiro-Villavicencio, H., 2021. Impact of BPA on behavior, neurodevelopment and neurodegeneration. Frontiers in bioscience, 26, pp.363-400.

Robertson, G.L., 2006. Safety and legislative aspects of packaging. Food packaging principles and practice, vol. 3. CRC Press, US, pp. 473–502.

Sadighara, P., Akbari, N., Mostashari, P., Yazdanfar, N. and Shokri, S., 2022. The amount and detection method of styrene in foods: a systematic review and meta-analysis. Food Chemistry: X, p.100238.

Samonsek, J., Puype, F., 2013. Occurrence of brominated flame retardants in black thermocups and selected kitchen utensils purchased on the European market. Food Addit. Contam. 30 (11), 1976–1986.

Schyns, Z.O. and Shaver, M.P., 2021. Mechanical recycling of packaging plastics: A review. Macromolecular rapid communications, 42(3), p.2000415.

Shallis, R.M. and Gore, S.D., 2022. Agent Orange and dioxin-induced myeloid leukemia: a weaponized vehicle of leukemogenesis. Leukemia & Lymphoma, pp.1-10.

Silano, V., Bolognesi, C., Castle, L., Cravedi, J.-P., Engel, K.-H., Fowler, P., Franz, R., Grob, K., Gürtler, R., Husøy, T., *et al.*, 2008. Note for Guidance For the Preparation of an Application for the Safety Assessment of a Substance to be used in Plastic Food Contact Materials. EFSA J. 6, 21.

Silva, A.S., Freire, C.J.M., García, R.S., Franz, R., Losada, P.P., 2007. Time-temperature study of the kinetics of migration of DPBD from plastic into chocolate, chocolate spread and margarine. Food Res. Int. 40, 679–686.

Skrzydlewska, E., Balcerzak, M., Vanhaecke, F., 2003. Determination of chromium, cadmium and lead in food-packaging materials by axial inductively coupled plasma time-of-flightmass spectrometry.

Su, Q.-Z.; Vera, P.; Van de Wiele, C.; Nerín, C.; Lin, Q.-B.; Zhong, H.-N. Non-target screening of (semi-)volatiles in food-grade polymers by comparison of atmospheric pressure gas chromatography quadrupole time-of-flight and electron ionization mass spectrometry. Talanta 2019, 202, 285–296.

Szynal, T., Rebeniak, M. and Mania, M., 2016. Migration studies of nickel and chromium from ceramic and glass tableware into food simulants. Roczniki Państwowego Zakładu Higieny, 67(3).

Tang, W., Hemm, I., Eisenbrand, G., 2000. Estimation of Human exposure to styrene and ethylbenzene. Toxicology. 144, 39–50.

Tawfik, M.S., Huyghebaert, A., 1998. Polystyrene cups and containers: styrenemigration. Food Addit. Contam. 15 (5), 592–599.

Taylor, S.R., 1964. Abundance of chemical elements in the continental crust: a new table. Geochim. Cosmochim. Acta. 28, 1273–1285.

Tighrine, A.; Amir, Y.; Alfaro, P.; Mamou, M.; Nerín, C. Simultaneous extraction and analysis of preservatives and artificial sweeteners in juices by salting out liquid-liquid extraction method prior to ultra-high performance liquid chromatography. Food Chem. 2019, 277, 586–594.

Triantafyllou, V.I., Akrida-Demertzi, K., Demertzi, P.G., 2007. A study on the migration of organic pollutants from recycled paperboard packaging materials to solid food matrices. Food Chem. 101 (4), 1759–1768.

Tsochatzis, E.D., Lopes, J.A. and Corredig, M., 2022. Chemical testing of mechanically recycled polyethylene terephthalate for food packaging in the European Union. Resources, Conservation and Recycling, 179, p.106096.

Vilarinho, F., Sendón, R., Van Der Kellen, A., Vaz, M.F. and Silva, A.S., 2019. Bisphenol A in food as a result of its migration from food packaging. Trends in food science & technology, 91, pp.33-65.

Wang, Yufei, and Haifeng Qian. "Phthalates and Their Impacts on Human Health." Healthcare (Basel, Switzerland) vol. 9,5 603. 18 May. 2021, doi:10.3390/healthcare9050603.

Wu, Y., 2014. General standard for contaminants and toxins in food and feed. International Food Standards.

Yam, K.L. and Lee, D.S. eds., 2012. Emerging food packaging technologies: principles and practice. Elsevier.

Yüzbasi, N., Sezgin, E., Yıldırım, M., Yıldırım, Z., 2003. Survey of lead, cadmium, iron, copper and zinc in Kasar cheese. Food Addit. Contam. 20, 464–469.

3

Food Packaging Materials and Safety Measures

Priti Khemariya and Tanweer Alam

Indian Institute of Packaging, Delhi

Food Contact Materials (FCM)

Before food consumption, during production, processing, preparation, packaging, packing, transportation, storage, serving etc. food comes in contact o many materials, all these materials are called food contact materials (FCM). FCM may be containers used for transportation for example milk container, machineries for its production or processing, packaging materials for its packaging like cans of soft drinks, kitchenware used for its cooking, serving, heating, washing. The selection of FCM requires intense care because FCM can affect human health and also the quality of food. During contact with food, molecules can migrate from FCM to food. Hence, FCM should have the following common attributes:

1. FCM should be compatible with the materials used for production, processing, transport, storage, serving etc.
2. FCM should be inert and non-reactive.
3. FCM should withstand the chemical, Mechanical and Thermal variations.
4. FCM should be clean, hygienic and tamper proof.
5. FCM should not have blisters, cracks, sharp edges and visible defects.
6. FCM should be free from rust.

Regulation for FCM

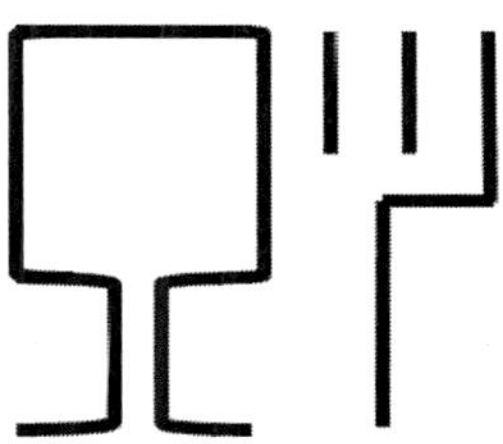

To ensure food safety to consumers regulation for FCM has been made. An international symbol " a wine glass and a fork" has been given for food safe materials which shows that the materials is safe when comes in contact with food and no health hazard will be caused. This regulation is applied to each material which comes under FCM. This symbol is applied in Europe, North America and some Asian parts.

Food Packaging Material

Food packaging material is one of the important food contact material. As it is well known that serving the safe food to consumers is a biggest challenge to our food manufacturing industries. Providing safe food to consumers gives following advantages to the food manufacturing industries:

1. It increases business survival percentage.
2. It increases the faith in brand.
3. It is an economic driver to industries.

Packaging plays an important role in serving safe foods to customers. With the urbanization and changing life style, food packaging industry has become very prompt to fulfil the demand of consumers.

- Good packaging protects the food from a safety and quality standpoint.
- Packaging helps in handling, storage, disposal, transportation and marketing of food.
- Packaging reduces food waste and preservatives used in food.
- Packaging protects food as biological barrier from microbiological contamination and pests, rodents etc. and hence preventing disease and spoilage.
- Packaging also controls rate of senescence i.e. ripening and ageing.
- Packaging protects food from shock or vibration like damage.
- Packaging acts as chemical barrier for oxygen, moisture and light.
- Packaging protects the package by tracking and tracing.
- Packaging provides products' information to customer by its labelling.
- Packaging minimizes the efforts necessary to prepare and serve food for example microwavable packaging helps consumers to cook food easily.
- Packaging provides some convenience features to consumers for example product can be visible from outside i.e. transparent packaging. Resealability helps to keep package for further use.

Therefore, selection of proper food packaging has to be done carefully as per the demand of food to be packed.

In food packaging industry, food is packed in various packaging materials.

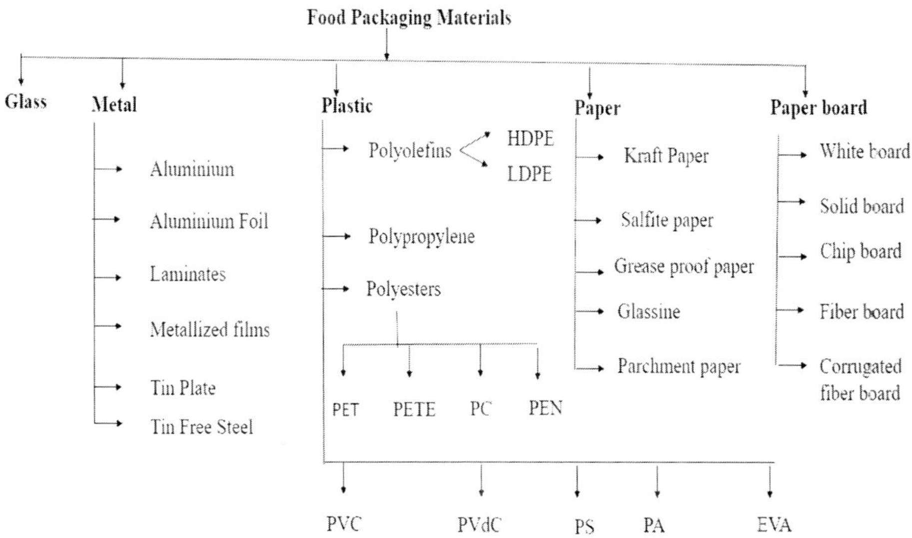

Glass

Glass is used as food packaging material because of following properties:

- Glass is odorless.
- It is non-reactive with food.
- It is impermeable to gases and vapors hence food remains fresh and tasty for long time period.
- It can withstand high thermal treatment so can be sterilized and hence can be used for sterilization of acidic food.
- It is rigid.
- It has insulation.
- It is transparent so consumer can see the food from outside without opening the package.
- It is reusable and recyclable hence is environmentally friendly packaging material.

Glass as packaging materials has the following limitations:

- It is heavy weight hence takes more transportation costs.
- It is very brittle in nature hence susceptible to breakage.

Metal

Among metals, Aluminium, Aluminium foil, laminates, metalized films tinplate, tin free steel are used as food packaging materials. All metallic packaging materials provides excellent physical protection and barrier property. These are recyclable hence are environmental friendly.

Aluminium

Aluminium is used as food packaging material because of following properties:

- It is light weight.
- It is corrosion resistant.
- It has barrier property to air, temperature, moisture and chemicals.
- It is flexible in nature.
- It is malleable.
- Example of aluminium as food packaging material is soft drink cans.

Aluminium as packaging materials has the following limitations:

- It is highly costly.
- It is unable to be welded so seamless containers are prepared from it.

Aluminium Foil

Aluminium foil is used as food packaging material because of following properties:

- It provides barrier property to air, moisture, odor, light, microorganisms etc.
- Acidic food can be packed.
- The thinner foil is used to wrap food.
- The thicker foil is used to make trays.

Laminates

It is a combination of Aluminium foil plus paper or plastic. Laminates is used as food packaging material because of following properties:

- It is heat sealable.
- It provides barrier to air and moisture.

Laminates as packaging materials has the following limitations:

- It is expensive so used to pack high value foods like spices, herbs.

Metallized Films

It is Aluminium containing plastic film. Metaliized film is used as food packaging material because of following properties:

- It provides better barrier property to air, moisture and odor than laminates.

- It is having better flexibility.
- It is used to pack snacks etc.

Tin Plate

Tin plate is used as food packaging material because of following properties:

- It has excellent barrier property to gases, water vapor, light and odor.
- It is heat resistant so sterile products can be packed inside.
- It is hermatically sealed.
- It is ductile so food containers of different shapes can be manufactured from it.
- It is low mol. Wt. and having high mechanical strength hence it makes transportation easy.
- It is used for making cans for drinks, processed foods.
- It is used to manufacture containers for powdered food like flour and sugar etc.

Tin Free Steel

It is also named as electrolytic chromium and chrome oxide coated steel. Tin free steel is used as food packaging material because of following properties:

- It is corrosion resistant.
- It is excellent for adhesion of coating like paints, lacquers and inks.
- It has high mechanical strength.
- It is less expensive than tinplate.
- It is used to manufacture food cans, trays, bottle caps, closure for retail packaging and drums for bulk packaging and storage.

Plastic

Plastic is used as food packaging material because of following properties:

- It is moldable material so used to make different sheets, shapes and structure.
- It is chemically resistant.
- It is inexpensive.
- It is light wt., so transportation is easy.
- It is heat sealable and microwavable.
- It is easy to print.

- Thermoplastics are used as food packaging material because they get soften upon exposure to heat and can easily be changed in different shapes like bottles, jugs, plastic films.
- It is low cost material.

Plastic as packaging materials has the following limitations:

- It has variable permeability to gases, vapors, light and low mol. Wt. molecules.

There are multiple types of plastics like polyolefin, polyester, PVC, PVdc, PS, PA, EVA are used as different types of food packaging materials.

a) **Polyolefins** Polyolefins is the collective word for polypropylene and polyethylene i.e.HDPE & LDPE.

HPDE (High Density Polyethylene)-

- HDPE is stiff, strong and tough material.
- It is resistant to chemical and moisture.
- It is permeable to gases.
- It is easy to manufacture and process.
- It is most recyclable plastic.
- It is used to manufacture milk bottles, juice bottles, water bottles, retail bags, grocery bags, trash bags etc.

LDPE (Low Density Polyethylene)-

- It is flexible, strong, tough and sealable.
- It is resistant to moisture.
- It is transparent plastic.
- It is used to make sealed food bags, squeezable food bottles, flexible lids etc.

b) **Polypropylene (PP)**

- It is harder, denser and transparent than PE.
- It is resistant to chemicals, water vapors.
- It can withstand 160°C temperature so is thermo-resistant.
- It is used in microwavable packaging and hot fill food packaging.
- It is mostly used to make yogurt container, margarine tubs etc.
- PP with EVA or PVdC provides strength and moisture barrier.eg. Salad dressing bottles.

c) Polyester (PE)

Among polyesters, PETE is more common than PET, PC and PEN.

PETE (Polyethylene terephthalate)

- It is light wt. and provides barrier to gases, moisture.
- It has good resistance to heat, mineral oil, solvents, acids but no bases.
- It is used to make beverages bottle, mineral water bottle, carbonated drink bottle, jars, tubs, bags, snack food wrapper.

PC (Polycarbonate)

- It is clear, heat resistant and durable plastic.
- It is used to make sterilizable baby bottles, refillable water bottles etc.
- The main limitation of this plastic is that it releases bisphenol A on washing with harsh detergent like Sodium hypochlorite.

PEN (Polyethylene naphthalate)

- It provides barrier to gases and water vapor.
- It is heat resistance ability.
- It protects against the transfer of flavors and odors.
- It is used for hot refills.
- It is used for manufacturing beverage bottles like beer.

d) Polyvinyl chloride (PVC)

- It is heavy, stiff, ductile, amorphous and transparent material.
- It is acid and base resistant.
- It has resistance ability against grease and oils.
- It is used for blister packaging like meat products packaging.
- It is difficult to recycle.
- Incineration releases chlorine which causes environmental problems.

e) Polyvinyldene chloride (PVdC)

- It is heat sealable material.
- It provides barrier to water vapor, gases, fat and oils.
- It is used for hot filling and MAP and retort packaging.
- It is used in flexible packaging of poultry meat, cured meat, cheese, snacks, confectionary, coffee, tea etc.
- Its recycling is difficult as it also releases chlorine on incineration.

f) Polystyrene (PS)

- It is clear, hard and brittle material.
- It has low melting point.
- It is used to manufacture protective packaging like egg carton.
- It is also used to make containers, cups, lids, plates, bottles trays etc.
- Its recycling or incineration is easy.

g) Polyamide (PA)

- It is also known as Nylon.
- It has mechanical strength and thermal property.
- It provides good chemical resistance, toughness and low gas permeability.
- It is used in boil-in bag packaging.

h) Ethylene vinyl alcohol (EVA)

- It provides barrier to oxygen, fat and oil.
- It is moisture sensitive. Hence, it is used where liquid is not in contact with packaging material.

Paper

- It has poor barrier property.
- It is not heat sealable.
- Its functional quality is enhanced by coating, laminating, impregnation with wax, resins, lacquers etc.
- Types of papers is as follows:

a) Kraft paper

- It is strong paper material.
- It is used for bags, wrapping of flour, sugar, dried horticultural crops.

b) Sulfite paper

- It is light wt. and weak paper material.
- It is used to make small bags, wrappers for confectionary, biscuits, cookies etc.
- Its glazing is done to increase wet strength and oil resistance.

c) Grease proof paper

- It is used to wrap snack food, cookies, bars, candy and oily products.

d) Glassine

- It has highly smooth, dense and glossy finish.
- It is used as liner for biscuits, butter, baked products, fast food etc.

e) Parchment paper

- It is not heat sealable.
- It doesn't provide good barrier to air and moisture.
- It is used to pack fat like butter.

Paper board

- It is thicker material than paper.
- It is used to make cartons, boxes, trays etc.
- It is used in shipping of goods.
- **Types of paper board is as follows:**

i. White board

- It is inner layer of carton.
- It is coated with wax or laminated with polyethylene.
- It is used to pack food directly.

ii. Solid board

- It has strength and is durable.
- When laminated with polyethylene, it is used to create liquid carton.
- It is used to pack milk, juice, soft drinks etc.

iii. Chip board

- It is recycled paper material so not used for direct food packaging.
- It is used to make outer layers of food cartons like tea, cereals etc.

iv. Fiber board

- Solid fiber board provides good barrier property.
- It is used to pack dry products like coffee, milk powder.

v. Corrugated fiber board

- It is resistant to impact abrasion and crushing damage.
- It is used to transport bulk or retail food package.

Food Hazard

Food is processed, partially processed or unprocessed substance which is used for consumption to get energy from it. Every consumer demands safe food or the food with zero risk to health. However, as per food manufacturer industry, food with zero risk is not at all feasible, hence, it is considered that food should have risk at an acceptable limit. But the question is what should be an acceptable limit. It is an issue because consumers have varying level of susceptibility and life style. Besides, unsafe food is that food constituents which affects consumers' health adversely. For example, pesticides, insects, worms, coloring agent, preservatives when present in food can cause adverse effect on consumers' health. Moreover, poor handling, processing and storage may also cause food hazards to consumers. The potential health hazards associated with food or potential risks to food safety can be categorized as follows:

1. **Physical Hazards of Food:** The hard foreign things which are found in food and cause physical injury or discomfort to consumers is considered as physical hazards. For examples, metal piece, glass piece, stones, soil particles, hairs, nails, bone piece, straw, feathers, excreta, iron piece, plastic piece, wooden piece, insects, paper piece, cardboard piece etc. The identification of physical hazards in food is easy and don't any expertise in its recognition.
2. **Chemical Hazards in Food:** Any toxic chemical which are found in food which may be added intentionally or unintentionally in food is termed as chemical hazard. For example: pesticides, drug, heavy metals, antibiotics, fertilizer, hormones, insecticides, lubricants, coatings, paints, refrigerants, water treatment chemicals etc. Some natural compounds which cause health hazards are also found in food. They are called as food allergens. For example, egg, milk, fish, egg, soybean, pea nut, wheat gluten, chemical and biological preservatives, additives, flavorings agents, colorants etc.
3. **Biological Hazards in Food:** Ingestion of food infected with pathogens and their toxic metabolites are responsible to cause biological hazards. This infection takes place during various steps in food. For example, harvesting, transport, storage, post-harvest processing, distribution and consumption. The consumption of infected food can cause food borne illness. The flora cause infection may include bacteria, yeast, mold, mycotoxin, viruses, protozoa, parasites etc. The common food borne illness caused by virus is hepatitis A, diarrhea in infants and young children, *Salmonella typhi, Salmonell paratyphi, Escherichia coli, Shigella dysenteriae, Vibrio cholera, Listeria monocytogenes, Helicobacter pylori, Compylobacter jejuni* are common food borne illness causing bacteria. Bacterial infection in food causes food infection and bacterial toxin ingestion causes food poisoning. Food poisoning. Food poisoning caused by toxin of *Clostridium botulinum, Clostridium perfringens* etc. are most common examples. Mycotoxin producing fungal species in pre and post-harvest crops, mycotoxin are produced due to the presence of fungal growth. These mycotoxin come to our

food chain. These even cannot be destroyed in thermal treatment of food thus during pre and post-harvest storage, prevention of fungal growth is necessary.

Some parasites are also common risk for food safety which come to our food due to the use of contaminated water or mishandling of food. For example, *Entamoeba histolytica* causes Amoebic dysentery. Giardia causes Giardiasis where diarrhea with green stools is the common symptom. *Toxoplasma gondii* causes Toxoplasmosis due to raw or undercooked meat consumption.

4. **Radiological Hazards in Food:** K-40 and Ra-226 are some naturally occurring radioactive element may be found in food. Similarly, some manmade radioactive elements such as Sr-90, Cs-137, I-131 may enter in our food due to nuclear accidents and this way can affect our health.

In view of the above hazards, to maintain the food safety is must to protect consumers' health. To provide food safety, food manufacturers should follow food safety management system. They should have to implement HACCP plan for safe food production. By following these, consumers can be protected from food borne illness.

HACCP

HACCP is Hazard Analysis Critical Control Point. HACCP is a plan to control risks to ensure food safety. It includes to limit the risks to an acceptable level or risks are eliminated completely.

HACCP includes following 7 principles to conduct-

1. Analysis of hazards
2. Identification of critical control points
3. Setting up critical limits
4. Setting up monitoring procedures
5. Conducting corrective action
6. Record keeping procedure
7. Verifying the documents

SQF

SQF is Safe Quality Food. Under the HACCP plan, SQF is food safety management system in which 3 certification of SQF level is maintained to assure customers about highest quality and safe packaging for the food products. SQF is recognized worldwide which promotes confidence for food safety and quality. It contains some integrated processes to control/eliminate/ minimize safety hazards in food packaging.

As per the HACCP plan and SQF certification, following are the requirements to maintain food safety or control/minimize risks.

1. Organizational employees role

- Organization should have a quality unit which approve or reject materials /procedures/specifications.
- All employees of the organization should have enough training, knowledge, experience and skills.
- All working staff should wear clean clothes and protective apparel for example gloves, hair net etc.
- All working staff should have good sanitation practice like washing hands properly.

2. Organizational infrastructural role

- Infrastructure should have enough space to store raw materials and finished products properly.
- Specific areas should be there for specific function for example discarding materials should keep separately in quarantine and finished products should be placed in separate area.
- Adequate environmental condition should be maintained for example light, temperature, humidity etc.
- Sewage and discarding materials should be disposed of by keeping the safety measures.

3. Machineries status

- All equipment should be in working mode. Their schedule of maintenance and cleaning should be written properly.
- ID number should be mentioned on each equipment.

4. Raw materials status

- Raw materials should be kept protected from contamination in specific area.
- Their receipt, handling and sampling method should be written and followed timely.
- Every lot of raw material should be named with code and lot number.

5. Production and process status

- Documentation of procedures is must.
- Contamination should not be raised.

- Work area should be cleaned.
- Procedure of production should be systematic.

6. Warehouse handling

- Warehouse should be cleaned and clearly identified with sections.
- There should be first in first out (FIFO) system.

7. Laboratory status

- Laboratory should have calibration and maintenance of each equipment used.
- Laboratory should have written specifications, Standards, sampling and test procedures.
- The laboratory and working procedures should be supervised properly.
- Sample should be reserved for a determined period of time.

8. Documentation

- Documentation of each section for example production, quality control training, maintenance, repairing, cleaning should be maintained.
- Each practice should be documented on paper with correct date and time.

The above all requirements come under CGGMP'S (Current Good Manufacturing Practices). By following all CGGMP'S, food safety can be maintained as per the HACCP plan and SQF certification can be availed to assure customers to packaging for food safety and quality.

Suggested References

- Marsh K and Bugusu B (2007), Food Packaging—Roles, Materials, and Environmental Issues 72 (3), Journal of Food Science (doi: 10.1111/j.1750-3841.2007.00301.x)

4

Packaging Design and Food Packaging Material

Priyanka Prajapati[1], Meenakshi Garg[2*], Rajni Chopra[3], and Susmita Dey Sadhu[4]

[1]Ph.D Scholar, Department of Food Technology, Bhaskaracharya College of Applied Sciences, University of Delhi, New Delhi
[2]Associate Prof., Department of Food Technology, Bhaskaracharya College of Applied Sciences, University of Delhi, New Delhi*
[3]Associate Prof., Department of Food Science and Technology, National Institute of Food Technology Entrepreneurship and Management, Sonipat, Haryana
[4]Associate Prof, Department of Polymer Science, Bhaskaracharya College of Applied Sciences, University of Delhi, New Delhi

Abstract

The packaging of commercial food products plays a crucial role in marketing, acting as a silent salesperson that conveys a product's image and brand value within the retail environment. A well-designed food package can stimulate a multi-sensory customer experience, influencing attention, perceived value, and understanding of product functionality. Ultimately, effective packaging can even drive consumption and significantly impact consumer buying decisions.

This chapter delves into the various design elements that influence consumer perception and purchasing decisions. These elements include both verbal cues (nutritional information, taglines) and visual attributes (color, shape, and illustrations). Additionally, the chapter explores the different packaging materials available and their physicochemical properties, all of which are important considerations during the food package design process.

Introduction

Packaging is an integral and pervasive part of contemporary world. Packaging makes material handling convenient, economical and effortless and modern consumer marketing will seem virtually impossible without packaging. For an engineer or food scientist, packaging is specialized arrangement of matrices fabricated for the delivery of safe food products with maintaining high quality and food safety, at every step of distribution chain including procurement of raw material to manufacturing of finished product; packaging of product; storage; transportation and distribution to wholesaler, retailer and consumer and even for recycling operations [17].

Packaging of food is imperative in a world dominated by marketing-based economy The fact that 73% of purchase decisions are made at point of sale highlighted the importance of a good package design in marketing. During scanning of food package, consumer made quick and rapid perception about the inclusion of a particular food based on the various verbal and nonverbal attributes embodied on food package [25]. Designing is an indispensable are of food packaging. It is medium to attract potential buyers and distinguished the product from the competitors. The presence of visual attributes on package design will not only ameliorates the food product and glamorizes the packets but also defines its authenticity and peculiarity. Therefore, a quality package design is characterized as communication channel between the manufacture and consumer, and unless performed correctly the market image of the brand and product suffers and influence consumer's perception.

Food Packaging Design

Apart from protection, communication with the consumer and convenience them to buy the product is one of the indispensable functions of food packaging [3]. Thus, package design has emerged intrusively in the sector of marketing diverse food products and is a prime advertiser to communicate the benefits to the customer. The importance of package design reflects from the fact that it influences consumers purchase decisions when they actively scan number of food packages of similar commodity placed in the same shelf of supermarket. It is categorized as an extremely influential medium because of its pervasive impact on consumers, and company brand value in the market. The design of packaging for a specific food commodity is within the hands of company and is influenced by complex set of factors from the business environment control of the company and are means of meeting changes within the business environment. Among the main influencing factors, emerging technology, material of fabrication, logistic requirements, environmental issues, Statutory Guideline, consumer preferences and business marketing aspects all are taken into consideration during designing of food package [13].

Package design has been describing as a broad term gauging technical attributes (e.g., ergonomics, mechanical skills, recyclability, barrier properties) and a package sensory attribute (Colour, shape) [15]. The most important aspect of package design is spanned around our sense of perceiving concinnity and space. The identification of specific target group identifies their requirement from the roots of package design. For example, when design a food package targeted at children, application of bright colors, cartoon characters, references to fun, sport and play, are the considering elements [16]. The packaging elements (color, shape, image, package material, logos and labelling influence the consumer behavior by making the marketing attractive and distinguishable among other products [3]. Ordinarily the elements present on a food package is classified between verbal (claims, logos, tag-line) and visual elements (color, graphic, shape, images) [19]. Consumer behavior of buying leisure food products is calibrated through three major factors: food attribute, brand scale, and price, which are all closely related to package color[3]. The main points that need to be focused while designing a food package are stated below

i. Nature or Type of Product: The nature of product will convey the ideas and concepts for designing the structure and visual element of style, mood and tone of package.

ii. Target Group: Identifying and targeting correct audience is one of the most important activities in marketing management.

iii. Identity Factor: The unique group of companies or brand related to the graphic design are the main key element which make the product distinguishable from its competitors.

iv. Element of Graphic design: To develop a visually appealing package the visual elements such as the layout of text, images, colors and size will be designed properly to arouse the interest of consumer.

v. Aesthetic Factor: The aesthetic factors or visual attributes should be placed in such a way that must create a visual sense of beauty, good impression, good psychological presence, and positive reactions.

vi. Added Value: Appeal is the main (Theme) of the advertising that will attract and convince the consumer to purchase the food product.

vii. Structure factor: The main concern that will take into account while designing food package is its structure that should provide adequate protection to food product.

viii. Form of packaging: The form of product packaging must be compatible with the food product type in order to aids in consumer convenience.

ix. Long Term Environmental or Legal Issues: Package development involves considerations for sustainability, environmental responsibility, and applicable environmental and recycling regulations.

x. Cost: The balance of marketing planning and morphology of budget cost and profits when entrepreneur's launching of the product and distribution is met.

Color design on Package Food

When it comes to packaging design, color have been positioned at a superiority. Color design is a keynote language that cross the obstacles of different languages, and cultural diversity, and aids in delivering information about commodity. As described in psychology, consumer will resonate and pay more attention to contrasting, bright colors of food package instead of its shape, pattern, and structure. Color design is the major visual language for package, and sets the keynote for the emotional design of leisure food package.[4] Opting appropriate color for food package design deliever number of benefits adding to the emotional resonance that consumers have for food package, stirring up consumers' desire to buy, thus promoting food sales, and shaping and communicating the brand image for the leisure food.

Major package design stakeholders like Pentawards and iF, give supreme priority to visual color that arose consumer desire for buying. Appropriate color matching for food package is necessary for increasing marketing scale. Researchers have postulated that more saturated and brighter color contrasting with surrounding are visually more attractive to consumer [14]. During purchasing leisure food consumer will first attracted by colors and make a subconscious judgment about the product, then pay attention to other information on food package (brand, health claims, nutritional information). Moreover, consumer associate some specific color to the particular type of food or brand. For example, brown color is associated with chocolate, pink color is strawberry and orange color is with orange flavor. The science of color design is meant to ameliorate the consumer buying experience and harmonize the consumer emotional satisfaction to the package color. Thus, the art of color design can propel the positive cycle of food buying [3].

Shape and Size

Adding convenience to package food is one of the major factors which will be considered while designing food package. One of the important aspects of convenience is apportionment function of packaging. Large scale manufacturing of food product creates the thrust for developing effective methods of apportioning the product into consumer-sized dimensions. For a food product that is not entirely consumed at the time of first opening the package, the package should offer the feature of resealing or repacking until the product is entirely consumed [8].

Shape of the primary package is another important aspect which aids in consumer convenience [8]. The shape of food package should fulfill basic functionalities like easy to use, open and pour as appropriate and have the ability to form the secondary and tertiary packages. In addition, the shape of food package is also a identity mark of specific brands such as the hourglass shape of the Coca-Cola bottle or the round-shouldered Bordeaux-style wine bottles [8]. Garber *et al.* 2009 describes major factors that affects the shape of food package and consumer buying behavior. Their study has reflected that apparent size of food package and not its actual size that affects the consumer purchase and consumption. Different shapes provide distinct effect on perceived volume appearance, geometrically simpler shapes are recognized as larger than geometrically complex shape. Food Package or Bottles with tall bodies are perceived as larger and voluminous as depicted in Fig :1 [5].

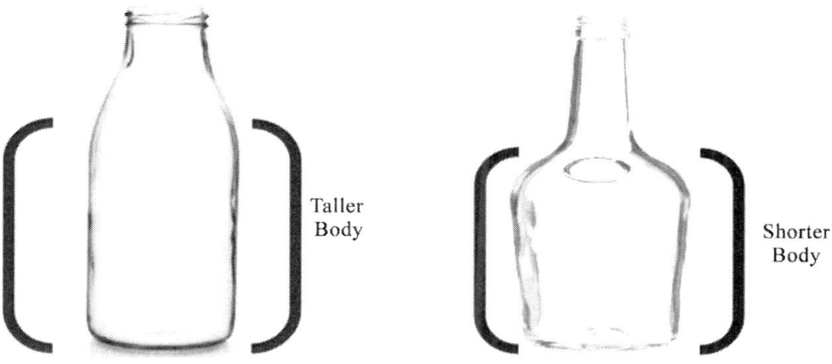

Figure 1: Different shape of food bottles with different body size

Labels and Logo

Brand labeling is the strategy designed by the organization to help the people to quickly identify and experience their brand, and also give them a reason to choose their products over the competition's or rival's product. Henderson and Cote (1998) generate insights that assist managers in selecting or modifying logo designs to achieve corporate image goals. They identified seven main factors for designing a logo of which three are of major importance. The first factor, "natural," combines lower-level characteristics, such as representative and organic, and reflects the degree to which constitutive design elements depict commonly experienced objects. The second factor, "harmony," combines symmetry and balance. A logo that uses design elements that are symmetric around both a horizontal and a vertical axis would be more harmonious than one that does not. The third factor, "elaborate," is a combination of design element complexity, activity, and depth.

Verbal Information

According to studies, food descriptions intended to appeal to multiple senses generate improved taste perceptions than single-sense descriptions. Because multiple senses (smell, sight, sound, and touch) collectively generate taste, descriptions mentioning these senses will be more influential than descriptions mentioning only taste. Information about the healthiness of the food also influences the food perception.[15]

Right visual placement of information creates more positive attitudes towards the product. Positioning information related to the tastiness (or other emotional elements) of a product are perceived more fluently when visually placed at the bottom of the packaging (vs. at the top). Conversely, information related to the rational elements of a product (e.g., its healthiness) are more fluently perceived when placed in the higher part of the packaging (vs. the lower).

Product Image

Aside from labels and product descriptions, packaging is usually composed of visual cues. The presence of a product image positively affects consumers' attitudes toward the package and their beliefs about sensory brand attributes. Likewise, a product image can boost consumer self-evaluations, and may increase the likelihood that consumers will use the image as an extrinsic cue and a product-quality indicator. The presence of a product picture can lead consumers to better imagine a product's taste, feel, smell, look, and sound. A product's picture, therefore, can set consumers' expectations and be an "advance organizer" for the other potentially available packaging information.

Package Structure Design Process

The package structure design process typically consists of different factors which helps in harmonize the needs of food manufacturers and consumers.

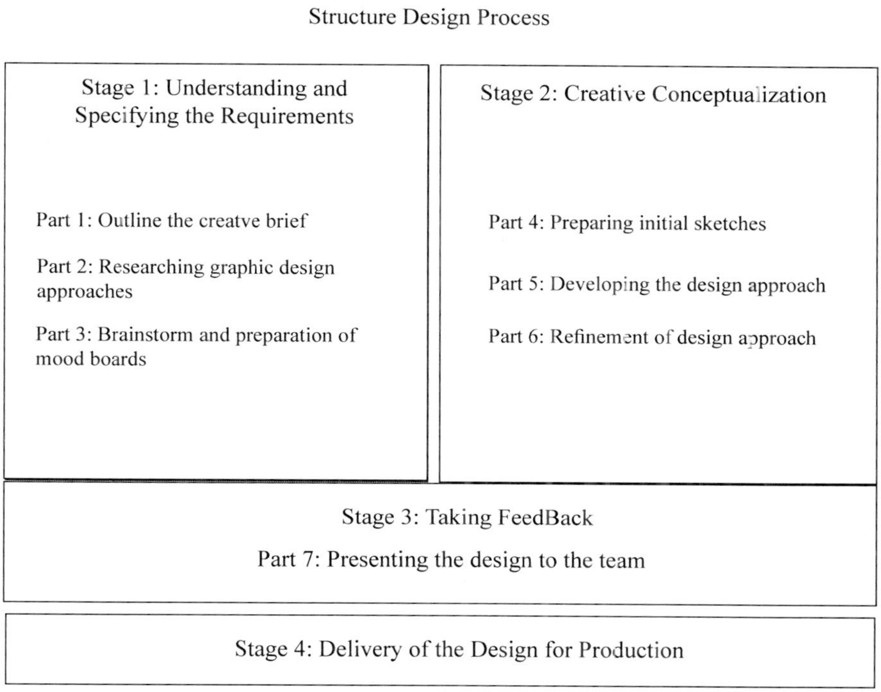

Figure 2: Illustration of Package designing Process

Package Design for Different Food Products

Package design is a set of keynote factors which influence the consumer's buying behavior. Package design is a reflection of brand's identity and a potent mode of communication with consumer. According to the theory postulate by Gestalt psychology package design is a holistic approach which signifies that cosumer's first impression about the food product is a combine response of the all elements present on the food package. The package design will vary according to the nature of food product. Table 1 provide an insight of different food products and their package design requirement.

Table 1: Characteristics of Food Package for different type of food.

S.No	Type of Food	Characteristics of Food Package	References
1.	Organic Food	Package material is composed of more paper material and more use of white and green colours, images displaying nature and organic claims.	Chrysochou & Festila 2019

2.	Milk	The package that has harmony in shape and background colour receives more attention from consumer. Presence of illustration like cow and buffalo explicit the origin of milk and linked it with naturalness Brand awareness and product history is an important factor that influence consumer's purchase decision.	Valajoozi & Zangi 2016
3.	Fruit Juices	The harmony of food package colour correlates with consumer perception of freshness and high quality. Brighter colour packs receives more attention by the consumer. Presence of fruit image on the front side of pack arouse the craving for juice among consumer. Presence of nutritional label and content on the back side also influence consumer's decision.	Wei *et al.* (2014). Nemergut, J., & Mokrý 2020
4.	Meat Products	Presence of label hold a supreme position while considering consumer buying pattern for meat. Organic label and veterinary stamps by government authorities affect consumer preference. "Halal" label is also important for Muslims which determines authenticity and wholesomeness.	Font-i-Furnols and Guerrero 2014
5.	Confectionery	The design appeal must be iconic and largely visual to attract young children	Saha *et al.* 2022

Graphic Design Significance

Graphic designing is an important aspect of package design which makes them visually distinguishable from other rival products. Package graphic design makes the first impression, founds the brand identity and foster brand loyalty. The main objective of package graphic design is to create effective visual impact, stimulate consumer interest and influence consumer purchase behavior[29]. The tools of package design, notably topography, image, layout, colour, typeface, patterns and ornaments constitute the brand identity and creates a strong impact on the brand image and customer perception about the food product. These observations construct the basis that graphical designing is a language of visuals which provide assistance to the marketers and designers to share the brand idea to consumers [27]. Designers and marketing members identify certain specific visual characteristic or "visual codes" for product and package design that signifies specific food categories. The category based visual designs refers to special shapes, color, typography and illustration which helps consumers to identify food products. For example French milk bottles available in the retail market are round or curve and have specific plain white color and blue screw cap and the same family of typefaces (bold, rounded, and sans serif lowercases) [28].This engrossed the customer interest and curiosity and finally reflects the gain in retail environment [27].

Table 2: Graphical design of different companies' logo

Sample	Brand	Typography	Color Scheme	Logo
Aerated Beverages	Coca Cola	Spencerian script		
Aerated Beverages	Pepsi-cola	Sans serif		
Instant Noodles	Maggie	Sans serif		
Instant Noodles	Top Ramen	Sans serif		

Food Packaging Material

The protection offered by a package is determined by the nature of the packaging material and the format or type of package construction. A wide variety of materials is used in packaging and primary packaging materials consist of one or more of the following materials: glass, metals, paper, and plastic polymers [8]. The selection of proper packaging materials and systems is an integral part of food process and product design [10]. Package design and construction play a significant role in determining the shelf life of a food product. The right selection of packaging materials and technologies maintains product quality and freshness during distribution and storage. Materials that have traditionally been used in food packaging include glass, metals (aluminum, foils and laminates, tinplate, and tin-free steel), paper and paperboards, and plastics. Moreover, a wider variety of plastics have been introduced in both rigid and flexible forms. Today's food packages often combine several materials to exploit each material's functional or aesthetic properties. Consumers demand packaging that causes less waste, incorporates recycled materials and is recyclable when empty.

Glass

Glass is an amorphous, inorganic product of fusion that has been cooled to a rigid condition without crystallizing. Although rigid, glass is a highly viscous liquid that exists in a vitreous or glassy state. The two main types of glass containers used in food packaging are bottles (which have narrow necks) and jars (which have wide mouths or openings). Approximately 75% of all glass food containers are bottles and approximately 80% of container glass is clear, the remainder being mainly amber or green. Today's glass containers are lighter but stronger than their predecessors, and through such developments the glass container has remained competitive and continues to play a significant although declining role in the packaging of foods. Although glass is an excellent barrier to water vapor, gases, and odors, an incorrectly designed or applied closure may negate the benefits that glass packaging offers in protecting foods from deterioration [8].

Figure 3: Different shapes of glass bottles

Plastics

Plastics are polymerized chain of carbon containing monomers linked by strong chemical forces. The characteristic feature of plastic is dependent on the type of monomer and its composition. These differences categorize the plastic into three different groups: thermosets, elastomers and thermoplastics. Thermoset plastics loose their remolding capacity after shaping. Thermoplastics founds various application in food packaging sector due to its recycling properties without altering their chemical and mechanical characteristics. Elastomers plastics as the name signifies has the property to regain their original shape after deformation. Some of the most widely used plastics are polypropylene (PP), polyethyleneterephtalate (PET) and polylactic acid (PLA). The enormous advantages of plastic such as printability, moldability, heat salability and makes it one of the extensively used packaging material in food sector. PET is widely used as replacement for glass in beverage bottles and for food packaging. PLA is the most common used biodegradable plastic alternative to fossil-based plastics such as PE[30]. The major drawbacks of plastics is their disparate barrier properties against UV light, moisture, odours , and low molecular weight chemicals [31]. Most concerning issue associated with plastic is their non degradable characteristics that opens the door for bioplastics. However plastic used in food packaging has continuously increased because of low cost and other functionalities such as optical stability, thermostability and microwavability.

Table 3: Different proprieties of plastics and its application in food

Type of Plastic Material	Barrier properties	Miscellaneous properties	Food product application
Low Density Polyethylene (LDPE)	Good moisture and oil resistant and very poor gas resistant	Tough, Flexible and has good sealing properties.	Breads and frozen food products, flexible food bottles
Linear Low Density Polyethylene (LLDP)	Good moisture and oil resistant and very poor gas resistant	Tough, Stretchable, provide barrier against grease and has good sealing properties.	Stretching wraps and heat seal-able films.
High Density Polyethylene (HDPE)	Excellent moisture resistant, oil resistant and very poor gas resistance	Tough, inelastic, strong, has good machinability and heat-sealing properties.	Liquid food products (milk, juice), margarine tubs, cereal box liners
Polypropylene (PP)	Good moisture resistant oil resistant and very poor gas resistance	Moderately hard and inelastic	Liquid food products (milk, juice), margarine tubs, cereal box liners and microwave packaging
Polyester	Good resistant to oil, moisture, gas and impact; low resistant to abrasion	Hard and inelastic	Refillable plastic bottles and baby bottles
Polyethylene terephthalate (PETE/PET)	Good resistant to oil, moisture and gas	Hard, stiff and strong	Bottles, jar tubs, tubes, trays, blisters and snack food wrappers

Polyethylene naphthalate	Good resistant to oil, moisture gas and UV light	Stiff and inelastic and preserve flavor of food products (wine and beer)	Wine and beer bottles; refilling and rewashing bottles
Polyvinyl Chloride (PVC)	Good resistant to oil, moisture and moderate resistant to oxygen	Hard, stiff, ductile and stable to electrical properties	Bottles and film; limited use in food packaging
Polyvinyl dichloride (PVDC)	Excellent moisture and gas resistant; good oil and grease resistant	Hard, stiff, ductile and stable to electrical properties	Cured meat, poultry, cheese, confectionery, snack food, tea and coffee
Polypropylene	Low moisture and gas resistant; moderate oil and grease resistant	Tough and brittle with low melting point	Protective packaging of eggs; disposable bottles, cups, plates and trays
Ethylene vinyl alcohol copolymer (EVOH)	Excellent moisture gas, oil and grease resistant	Hard, Stiff and strong	Used in coextruded film to provide barrier against water
Polyamide	Excellent moisture gas, oil and grease resistant	Hard and stiff	Used in boil in bags

Reference : Wani, A.A., Singh, P. and Langwoski,H.C. Food Packaging Materials: Testing & Quality Assurance. (2017). United Kingdom: CRC Press.

Metal

Primary four metals are commonly employed in packaging of food such as steel, aluminum, tin, and chromium [8]. Aluminum is widely used in food packaging to manufacture cans, foil, and laminated paper or plastic packaging. It is a lightweight, silvery white metal extracted from bauxite ore, where it exists in compound state with oxygen. Strength properties of Aluminum like ductability is enhanced by adding Magnesium and manganese. Wide application of aluminum in food products is due to its highly resistant properties to most forms of corrosion. The natural coating of aluminum formed after reacting with oxygen on the surface provides efficient barrier against air, temperature, moisture, and chemical attack. Additional advantages like excellent malleability and surface resilience and good formability makes the aluminum a perfect raw material to fabricate versatile food package. It is also an ideal material for recycling because it is easy to reclaim and process into new products. Pure aluminum is used for light packaging of primarily soft-drink cans, pet food, seafood, and pretreated closures.

Tin and steel, and chromium and steel, are used as composite materials in the form of tinplate and electrolytically chromium-coated steel (ECCS). Although tin provides steel with some corrosion resistance, tinplate containers are often lacquered to provide an inert barrier between the metal and the food product. Commonly used lacquers are materials in the epoxy phenolic and oleo resinous groups and vinyl resins. In addition to its excellent barrier properties to gases, water vapor, light, and odors, tinplate can be heat-treated and sealed hermetically, making it suitable for sterile products. Because it has good ductility and formability, tinplate can be used for containers of many different shapes. Thus, tinplate is widely used to form cans for drinks, processed foods, and aerosols; containers for powdered foods and sugar- or flour-based confections; and as package closures. Tinplate is an excellent substrate for modern metal coating and lithoprinting technology, enabling outstanding graphical decoration.

Paper and Paperboard

Paper and paperboards are versatile packaging material with wide range of uses and applications. Conventionally the are manufactured to form corrugated boxes, milk cartons, folding cartons, bags and sacks, and wrapping paper. The major advantages offered by Paper and paperboards are good mechanical strength, biodegradability and good printability. Coatings of plastics and waxes have been applied to paper in order to improve their poor barrier properties. Apart from their poor resistance properties other major disadvantages such as opacity, and poor heat salability limit their application to some food products [31].

Table 5: Different proprieties of Papers/paperboard and its application in food

Type of Paper	Food Product Application	Functional Properties
Kraft Paper	Sugar, Flour, Dried fruits and vegetables	It is the strongest form of paper available for food packaging. It found major application as bags and outer covering.
Sulfate Paper	Confectionery and biscuits	It is glazed to improve water and oil resistance. Provide an appropriate surface for printing thus used in laminates of plastic and metal.
Glassine	Confectionery, Biscuit, fats and oils, fast food and baked foods	Extreme hydration makes it dense, glossy and highly finished
Greaseproof	Snack foods, Cookies, Candy Bars, and other oily foods,	It has oil resistant property but permeable to water.
Parchment Paper	Butter and lard	Impervious to fat but not provide barrier against moisture and gases.

Chipboard	Outer layer of tea and cereals	Least expensive type of paper manufactured from recycled parts. It is unsuitable for direct contact with food, printing and folding
Paperboard	Bulk packaging of food products and case pack for retail products.	It provides good Physical barrier against compression, impact, abrasions and crushing.

Guidelines for packaging design

Previously, food manufacturers had flexibility in producing, packaging, labeling, and selling food due to a number of ambiguous standards. However, food packaging regulations are constantly evolving, and keeping pace with them is essential for manufacturers when designing food packaging.

To address the ambiguity of regulations, the Food Safety and Standards Authority of India (FSSAI) has established mandatory guidelines for food packaging and labeling, enforced under the Food Safety and Standards (Labelling and Display) Regulations, 2020. These regulations ensure transparency and consumer protection by mandating clear and accurate information on food packaging. The FSS labelling and display regulations specify essential information that must be displayed on food packaging, which includes:

a) **Product Name:** The clear and prominent identification of the food item.

b) **List of Ingredients:** A complete list of ingredients in descending order of weight or volume at the time of their incorporation into the food.

c) **Nutritional Information:** A standardized nutritional information panel detailing key nutrients like fat, saturated fat, sugars, salt, and others.

d) statement indicating whether the food is vegetarian or non-vegetarian.

e) **Declaration of Allergens:** Any food allergens or ingredients likely to cause hypersensitivity must be highlighted.

f) **Date of Manufacturing, Expiry Date, and Batch/Lot Identification:** Clear markings specifying the date of manufacture, expiry date, and batch/lot identification for traceability purposes.

g) **Name and Complete Address of the Manufacturer / Packer / Marketeer:** Identification of the entity responsible for the food product.

h) **Consumer Care Information:** Contact details for the manufacturer or importer for customer inquiries.

Indian Nutrition Rating (INR): This proposed system assigns a star rating (from 1 to 5) based on the overall nutritional profile of the food product. Higher ratings indicate healthier choices. This system would notify and compare the levels of fat, saturated fat, sugars, and salt in a food product using color codes. Red would indicate high fat, sugar, or salt content in packaged foods.

Future Trend

The packaging industry faces growing scrutiny from consumers, environmental protectors, activists, and the media. In response to consumer concerns and evolving regulations, brands are increasingly prioritizing recyclable or biodegradable packaging solutions. This trend is driving exploration of eco-friendly alternatives to plastic in food packaging design.

To keep pace with changing consumer preferences, food packaging designers are developing innovative ideas that positively influence buying decisions. Notable advancements likely to shape future package design include printed electronic circuits that illuminate upon contact, smart tags for improved shelf-life management, and enhanced tamper protection.

References

1. Adesegun Kehinde, B., Majid, I., Hussain, S., & Nanda, V. (2020). Innovations and future trends in product development and packaging technologies. Functional and Preservative Properties of Phytochemicals, 377–409. doi:10.1016/b978-0-12-818593-3.00013-0
2. Arrúa, A., Curutchet, M. R., Rey, N., Barreto, P., Golovchenko, N., Sellanes, A., ... & Ares, G. (2017). Impact of front-of-pack nutrition information and label design on children's choice of two snack foods: Comparison of warnings and the traffic-light system. Appetite, 116, 139-146.
3. Auttarapong, D. (2012). Package Design Expert System Based on Relation between Packaging and Perception of Customer. Procedia Engineering, 32, 307-314.
4. Berk, Z. (2013). Food Packaging. Food Process Engineering and Technology, 621–636. doi:10.1016/b978-0-12-415923-5.00027-7
5. Bloch, P. H. (1995). Seeking the ideal form: Product design and consumer response. Journal of marketing, 59(3), 16-29.
6. Celhay, F., & Trinquecoste, J.F. (2015). Package Graphic Design: Investigating the Variables that Moderate Consumer Response to Atypical Designs. Journal of Product Innovation Management, 32, 1014-1032.
7. Celhay, F., Cheng, P., Masson, J., & Li, W. (2020). Package graphic design and communication across cultures: An investigation of Chinese consumers' interpretation of imported wine labels. International Journal of Research in Marketing, 37, 108-128.
8. Cenci-Goga, B.T., Iulietto, M.F., Sechi, P., Borgogni, E., Karama, M., & Grispoldi, L. (2020). New Trends in Meat Packaging. Microbiology Research.
9. Chrysochou, P., & Festila, A. (2019). A content analysis of organic product package designs. Journal of Consumer Marketing.
10. de Matos Fonseca*, J., Koop*, B. L., Trevisol, T. C., Capello, C., Monteiro, A. R., & Valencia, G. A. (2022). An Overview of Biopolymers in Food Packaging Systems. Nanotechnology-Enhanced Food Packaging, 19-53.
11. Font-i-Furnols, M., & Guerrero, L. (2014). Consumer preference, behavior and perception about meat and meat products: an overview. Meat science, 98 3, 361-71 .
12. Food Packaging Materials: Testing & Quality Assurance. (2017). United Kingdom: CRC Press.

13. Garber, L.L., Hyatt, E.M., & Boya, Ü.Ö. (2009). The Effect of Package Shape on Apparent Volume: An Exploratory Study with Implications for Package Design. Journal of Marketing Theory and Practice, 17, 215 - 234.
14. Gupta, R. K., & Dudeja, P. (2017). Food packaging. Food Safety in the 21st Century, 547–553. doi:10.1016/b978-0-12-801773-9.00046-7
15. Krishna, A., Cian, L., & Aydnolu, N. Z. (2017). Sensory aspects of package design. Journal of Retailing, 93(1), 43-54.
16. Mahor, N. (2020). "Factor effect the customer buying behaviour by influence of packaging and labeling of product".
17. Marsh, K., & Bugusu, B. (2007). Food packaging—roles, materials, and environmental issues. Journal of food science, 72(3), R39-R55.
18. Mazhar, Mahera and Daud, Sayeda and Arz Bhutto, Sana and Mubeen, Muhammad, Impact of Product Packaging on Consumers Buying Behavior: Evidence from Karachi (April 2, 2015). Journal of Marketing and Consumer Research, Vol. 16, pp. 35-42, 2015, Available at SSRN: https://ssrn.com/abstract=2910712
19. Nemergut, J., & Mokrý, S. (2020). Influence of packaging attributes on perception of juice: Eye-tracking study.
20. Oladumiye, E.B. (2018). Graphic Design Theory Research and Application in Packaging Technology.
21. Orth, U. R., & Malkewitz, K. (2008). Holistic package design and consumer brand impressions. Journal of marketing, 72(3), 64-81.
22. Otto, S., Strenger, M., Maier-Nöth, A., & Schmid, M. (2021). Food packaging and sustainability – Consumer perception vs. correlated scientific facts: A review. Journal of Cleaner Production, 298, 126733.
23. Rettie, R.M., & Brewer, C. (2000). The verbal and visual components of package design. Journal of Product & Brand Management, 9, 56-70.
24. Robertson, G. L. (2005). Food packaging: principles and practice. CRC press.
25. Robertson, G. L. (2014). Food Packaging. Encyclopedia of Agriculture and Food Systems, 232–249. doi:10.1016/b978-0-444-52512-3.00063-2
26. Rundh, B. (2009). Packaging design: creating competitive advantage with product packaging. British Food Journal, 111, 988-1002.
27. Underwood, R.L., Klein, N.M. and Burke, R.R. (2001), "Packaging communication: attentional effects of product imagery", Journal of Product & Brand Management, Vol. 10 No. 7,pp. 403-22
28. Valajoozi, M. R., & Zangi, N. O. (2016). A review on visual criteria of pure milk packaging for parents and their children (case study: Tehran, Iran). British Food Journal.
29. Wei, S., Ou, L., Luo, M.R., & Hutchings, J.B. (2014). Package Design: Colour Harmony and Consumer Expectations. International Journal of Design, 8.
30. Wu, T., Li, Y., & Liu, Y. (2017). Study of Color Emotion Impact on Leisure Food Package Design. HCI.
31. Zhang, S. (2013). Efficient Color Design Scheme of Children Food Packaging.

5

Potential Hazards to Food Safety

Chinthu Udayarajan[1], Treeza Annie Jose[1], Supriya Shanmugam[2A], Sonika Kumari[2B] and Sanu Jacob[2C]*

[1]CSIR – National Institute for Interdisciplinary Science and Technology (NIIST) Thiruvananthapuram, Kerala, India
[2]Food Safety and Standards Authority of India (FSSAI), [2A]Bengaluru; [2B]New Delhi and [2C]Chennai, India

Introduction

The purpose of this e-book is to outline the main hazards in food production and processing that could influence the safety of foods and thus the health and wellbeing of consumers. The suggestive actions to restrain the formation of compounds or organisms that may threaten the safety of a food will be discussed along with the legal and regulatory framework.

"Safe Food Now for a Healthy Tomorrow"

Food is the third most basic human need, after air and water. Only when food is safe can it contribute to our food security and health. When food is not safe, people cannot thrive, hunger and poverty cannot be alleviated, and a healthy life is not possible. Food safety has a significant impact on human health, and food, which is essential for human survival, comes from plants and animals. Food safety is a core tenet of One Health.

This chapter gives an overview of the main hazards such as physical hazards, chemical hazards, biological hazards, and adulterants in different types of foods and how the hazards influence the safety of the food along with health and well-being of the consumers. The effective way of management of these hazards and different hazard limits set by the regulators such as FSSAI, USFDA, EU AND CODEX along with labelling requirements for the hazards has been discussed in this chapter.

Definitions of Food, Nutrition and their Importance for Well-being

The Food Safety and Standards Act (2006) defines Food as "any substance, whether processed, partially processed or unprocessed, which is intended for human consumption and includes primary food, genetically modified or engineered food or food containing such ingredients, infant food, packaged drinking water, alcoholic drink, chewing gum, and any substance, including water used into the food during its manufacture, preparation or treatment but does not include any animal feed, live animals unless they are prepared or processed for placing on the market for human consumption, plants, prior to harvesting, drugs and medicinal products, cosmetics, narcotic or psychotropic substances".

On the other hand, the World Health Organisation describes "Nutrition" as a critical part of health and development. Better nutrition is related to improved maternal and child health, stronger immune systems, safer pregnancy and childbirth, lower risk of non-communicable diseases and longevity. Health and nutrition are emphasized as critical components of basic needs in developing countries. Nutrition is a major determinant of health and productivity. Eating safe and wholesome food helps to overcome malnutrition.

Food Safety Challenges

Food safety is a scientific discipline, process or action that prevents food from containing substances that could harm human health. The objective of food safety is to have food that is safe to eat. The food should be free from toxic substances produced during contamination, adulteration, food poisoning / food intoxication and hazards such as physical, chemical, and microbiological that could enter during production, processing, transportation, storage, preparation or packaging of food products. One of the major global concerns that have a direct bearing on human health and survival is food safety.

Hazards: Contamination and Adulteration

Food contamination refers to a condition in which harmful microorganisms and chemicals are present in food that may cause illness to consumers. Generally, we conclude that the food is contaminated based on the changes in its flavor, texture, and appearance. These contaminants find their way into the food during various stages of production, treatment, packaging, or transportation. Since contamination has many ill effects on human health, many safety measures have been taken by the EU on reducing the number of contaminants within food products[1].

Adulteration can be defined as intentional (to increase the sale of a product) or unintentional removal of some value-added ingredient from the main food item with some inferior foreign particle which leads to lowering of the food quality[2]. Food adulteration is mainly associated with high-value export products like spices, herbs, oils, etc. Thus, to protect human health and for smooth trade purposes, World Trade Organisation (WTO) has laid some rules for Technical Barriers to Trade (TBT) and Sanitary or Phyto-Sanitary measures (SPS) agreements. These agreements help the member countries to understand and recognize each other's compliance assessment rules based on International Standards thus leading to smooth trade of products[3].

Different types of Food Contaminations

Based on various sources, food contamination is categorized into physical, chemical, and biological contamination.

Physical contamination: Physical contamination can occur at any stage of manufacturing, processing, and packaging. Most contaminants can harm and hold harmful bacteria. They wouldn't harm or degrade the food, but their presence in food can create health hazards for the consumer.

Chemical contamination: Chemical contamination paves way for a greater risk for human health. The chemical contaminants mainly include pesticides, cleaning chemicals, presence of heavy metals and so on. These contaminants if not controlled, can lead to chronic poisoning in humans.

Biological contamination: Biological contamination mainly happens due to the presence of bacteria or other microorganisms in food which causes the spoilage of food. The consumption of such spoiled food may lead to serious health issues[4].

Physical Contamination

Physical contamination of food happens when particles like hair, dirt, glass pieces or insects are present in food. Usually, physical contamination occurs during the stage of production or packaging of the food and is visible. Physical contamination is very distressing for the consumer. The presence of particles like nails or glass pieces in food leads to cuts in the mouth, breaking off a tooth or in some cases choking when swallowed. The presence of insect droppings or dead insects can lead to various diseases[3,4].

Pests: Rodents and Insects-their hairs or droppings

Pests like rodents and insects coming in contact with the food could cause a high risk of food poisoning as they are carriers of various disease-causing pathogens and microorganisms.

Rodents: Rodents mainly rats, and mice compete globally with humans for food, mainly causing pre-harvest damage to cereals. Rodent-borne diseases can reach humans directly when bitten by or through consuming food products that are contaminated with rodent faeces or urine or by inhaling germs that are present in their excrements.

Insects: Insects like flour beetles, cockroaches or termites bore into the foods like cereals, pulses and flours when kept in warm and humid conditions. They practically eat up the whole grain and leave their residues behind which have an offensive odour[3,4].

Packaging: Micro-plastics and inks

The migration of chemicals from food packaging materials to food is a major concern due to their ill-effects on human health. When the food is packed by a multi-layer of packaging material, migration can take place not only from the food contact surface but also from interior parts by diffusion. Plastics by virtue of low cost, versatility and lightweight nature are widely used as packaging material. They help in increasing the shelf life of the food as well as aid in transportation. Micro-plastics are smaller fragments of plastic mainly in the range of 5 μm which are produced when plastic is exposed to continuous sunlight, temperature changes and mechanical abrasions. The plastic packaging releases the micro-plastic during its daily use and thus the micro-plastic gets absorbed into the food we consume. Once ingested, micro-plastics have been found to mainly affect the immune system. Micro-plastics are generated daily, such as when we unwrap chocolates to eat, tear plastic tapes to open packed food, open a bottle to drink water etc.[5] Another commonly used material for packing dry foodstuffs like flours, sugar, vegetables, cereals etc. is paper, from which the migration of printing ink into the foodstuff is a major concern. The printing inks which are applied on the outside of the packaging find their way into the food by migration during the production and storage of food[6].

Physical Materials: Glass, Screw, Nails, Stones, Wood Pieces, etc.

Food contamination by physical material is unintentional; it usually happens during the food processing or packing stage. The main sources of physical contamination include presence of broken pieces of glass, screw, stone, wood pieces etc. in the food. These materials won't contaminate the food but if swallowed they might cause cuts to the lips or throat, breaking of the tooth or choking[3].

Chemical Contamination

It refers to a state in which there is presence of chemicals in an amount higher than the permissible limit that is considered as safe for consumption. The origin of chemical contaminants is mainly from polluted soil, air, water, pesticide residues, packaging materials and so on. Recently chemical food contamination has become a serious problem as the number of industries has escalated leading to environmental pollution. The plants that grow in polluted soil and the animals that consume polluted water pave the way to the entry of chemicals into the food chain. Nowadays, food is mixed with organic pesticides like DDT/ BHC, but these too leave a residue which on continuous consumption can lead to serious health issues. Food can also be contaminated by chemicals from the metals like zinc (used in galvanization). The storage of acidic food in unlined tin cans can cause certain reactions and get absorbed into the food. The ingestion of food contaminated with heavy metals can lead to gastrointestinal diseases[3,4].

Cleaning Chemicals

Cleaning during the processing of food helps in the removal of possible pathogens and is important for reducing food contamination. Thus, the chemicals used for cleaning should be of no impact on utensils used for the preparation of food and must be accepted by the law. The uses of sanitizers above permitted levels leave residues on fruits and vegetables. The chemical residues from cleaning liquids and disinfectants used for the cleaning of food handling utensils and its diffusion to food have been a major concern. Thus, the development of safe, nature-friendly cleaning formula is a major concern for food manufacturers. The use of food-grade and edible cleaning formula allowed by the FDA, ozone, inorganic peroxides were some of the alternatives to the chemical disinfectants. Recently technologies like the use of electrical pulses, beta and gamma irradiation for cleaning find interest among the food industries. In the case of these technologies, their impact on food stability and the formation of any side reagents should be studied[6].

Pesticides

Pesticides are agrochemicals used to destroy or repel any pest like insects, weeds etc. from destroying the crops. It is being used on a large scale to increase food production to meet the requirements of the rapidly growing human population. The use of fertilizers has increased the quality and quantity of fruits and vegetables but at the same time, it has caused adverse health effects. By the end of the 19th century, usage of organic fertilizers like DDT was started but the mineral phosphate fertilizers took over by the beginning of the 20th century. The studies showed that DDT has an immunosuppressive effect on human cells and prolonged intake into the body leads to various types of cancer. Pesticides are xenobiotic substances, which can alter the biological cycles and various chains in the ecosystem. Most of these pesticides are indestructible in nature and their existence in nature leads to biomagnification which is the accumulation of toxins at higher levels in the food chain. Pesticides can be categorized into various chemical classes:

Organochlorine pesticides: This category of pesticides belong to the group of chlorinated hydrocarbon derivatives. These pesticides include Dicofol,

Eldrin, Lindane, DDT, BHC, DDD, Chlorobenziate, Isodrin, Isobenzen, Methyloxychloro Aldrin, Chlorine, Heptaclor, Endosufan, Toxaphene and Chorapropylate are some of the organochlorine pesticides[7]. Organochlorines behave as endocrine disruptive substances by interfering with the molecular structure and operation of the endocrine system [8]. Numerous organochlorine compounds are neurotoxic and carcinogenic[9]. Gall stone disorders in humans have been linked to exposure to organochlorine pesticide residues[10].

Organophosphorus pesticides: It is the most used pesticide that is highly toxic against pests but doesn't cause harm to non-target species. It is widely used since it can be used against a variety of insects at the same time and stays in the environment for a very short period. However, recent reports showed that increased levels of organophosphates could be carcinogenic, mutagenic, and cytotoxic.

Carbamate Pesticides: Carbamates are cholinesterase inhibitors that cause build-up of acetylcholine in the nervous system which may affect sensory and behavioral characters.

Pyrethroid Pesticides: Pyrethroid pesticides are synthesized from the oleoresin of dried chrysanthemum. These pesticides are often combined with other chemicals to increase their longevity in the environment[11].

Heavy Metal: As, Pb, Hg, Methyl-Hg, Cd, Sn, Cu, Al, Cr, Ni, Se, Sb

Heavy metal contamination is a great threat to the human population and has enhanced due to industrialization and urbanization. These heavy metals find their way into the food through irrigation using the wastewater from industries, usage of chemical fertilizers and insecticides and other anthropogenic activities like mining. The accumulation of these metals leads to many abnormalities in humans that may prevail through generations[3,8].

Arsenic (As): Arsenic finds its way into the food mainly through the arsenic containing pesticides used in agriculture through mining activities and gets naturally released into Nature through volcanic eruptions etc. The intake of very small amounts even less than 5mg causes vomiting and diarrhea. Higher doses of arsenic may lead to acute poisoning.

Lead (Pb): Lead is a naturally occurring element and finds its way into the human body not only through food but also through contaminated water and dusted air. Various foodstuffs like shellfish, pectin, synthetic dyes, tea, baking powder, peaches etc. are sources of lead. The accumulation of lead in humans mainly affects the three systems viz. nervous system, renal system and hemopoietic system. Higher concentrations of lead in blood may lead to paralysis, poisoning, and behavioral disorders.

Mercury (Hg) & Methyl mercury (Methyl-Hg): Mercury is the only metal that is found in the liquid form, and it comes in contact with humans mainly through inhalation, injection, and through contact with skin. Exposure to mercury can cause serious mental disorders and can affect infants in the early stages of development. We mostly come in contact with methyl mercury by consuming fishes that are found in contaminated seas and oceans. Minamata, which is a neurological disease is mainly caused by the intake of mercury.

Cadmium (Cd): Cadmium is a white metal that is used in the coating of utensils and cookers. It gets dissolved when it comes in contact with acids present in fruits, juices etc. The presence of cadmium in the body causes health effects like hypertension, kidney lesions and can affect the calcium metabolism in the body. The Itai-Itai disease is due to cadmium poisoning which causes severe pain in the spine and joints.

Tin (Sn): The foodstuffs come in contact with tin during the process of packaging. The major sources of tin poisoning include storage of acidic foods like fruit juices in cans or leaving moist food in opened cans. A higher concentration of tin in the body causes abdominal pain, retaining of urine, dehydration, and severe headaches.

Copper (Cu): Copper finds its way into the human body when acidic foods are cooked in copper pans when copper is used in household plumbing. Acute poisoning by copper may lead to constrictions on the throat, diarrhea, etc[12].

Aluminum (Al): The consumption of aluminum occurs via the use of food additives, migration from the saucepan or food packing. It mainly affects the digestive tract and affects the normal functioning of the body[13].

Chromium (Cr)**:** Chromium is a toxic metal that occurs in Nature in various oxidation states. It is mainly released from mining and textile industries through the burning of coal and oil etc. The presence of chromium in the body causes kidney failures, abdominal pain and in severe cases, it may lead to death[3,8].

Nickel (Ni): The main sources of nickel include mining, burning of fossil fuels, release from automobiles, various metallurgical activities, and application of fertilizers. Contact with nickel causes cardiovascular diseases, asthma, respiratory tract cancer etc.[14]

Selenium (Se)**:** Selenium is a toxic metal that is mainly released into the water bodies from the industries as an effluent. It finds its way into humans through the food chain. The presence of selenium mainly affects the reproductive system in humans and a higher dose even leads to death[15].

Antimony (Sb): The main sources of antimony release into the environment include the textile and plastic industries, automobile emissions, bottled water and tinned foods. Short-term exposure to antimony in humans' results in irritation to the skin and eyes. Long-term exposure mainly causes inflammation of the lungs, chronic bronchitis, and chronic emphysema[16].

Mycotoxins

Mycotoxins are compounds produced by certain species of molds or fungi. They are toxic in nature and harmfully affect one quarter of the food produced in the world[17]. Frequent ingestion of mycotoxins may cause cancer. Common mycotoxins, crops affected, source and effect on humans are given in Table 1. Among the mycotoxins, aflatoxins are of the most concern.

Table 1: Common mycotoxins, commodity affected, sources and health effects

Mycotoxins	Commodities	Fungal Source	Effects of Ingestion
Aflatoxin B1, B2, G1, G2	Corn, Peanuts	*Aspergillus flavus* *Aspergillus parasiticus*	Aflatoxin B1 identified as potent human carcinogen by IARC[1]. Risk of human toxicosis
Deoxynivalenol Nivalenol Vomitoxin	Wheat, Corn, and Barley	*Fusarium graminearum* *Fusarium crookwellense* *Fusarium culmorum*	Human toxicoses in India, China, Japan
Zearalenone	Corn, wheat	*Fusarium graminearum* *Fusarium culmorum* *Fusarium crookwellense*	Identified by IARC as a potent carcinogen. Affects reproductive system
Ochratoxin A	Barley, wheat	*Aspergillus ochraceus* *Penicillium verrucosum*	Suspected by IARC as a human carcinogen
Fumonisin B1	Corn	*Fusarium moniliforme*	Suspected by IARC as a human carcinogen
Patulin	Apple juice	*Penicillium expansum*	Liver and kidney damages

[1]International Agency for Research on Cancer.
Source: Dohlman, 2003[17]

The climatic conditions in the tropical regions of the world such as high humidity, high temperatures and heavy rains during harvest etc. result in the proliferation of molds and the production of mycotoxins. Poor agricultural practices such as inappropriate methods of harvest, storage, transport, and sales also contribute to the increase in mycotoxin levels. Hence methods for abating the risk associated with mycotoxins is the implementation of good agricultural practices (GAPs) and good manufacturing practices (GMPs) at various points of food harvesting and processing. To control the risk of mycotoxin containing products reaching the consumers, product standards for Aflatoxins have been adopted by various organizations and Countries as shown in Table 2.

Table 2: Product standards for Aflatoxins (United States, European Union and Codex Alimentarius)

United States[1]		**European Union**[2]	
Product	**Standard**	**Product**	**Standard**
	Parts per billion		*Part per billion*
Raw Peanuts (industry standard)	15	Peanuts, nuts, dried fruit, and processed products there of, intended for direct human consumption	4 (3)
Human food, corn, grains for immature animals including poultry and dairy animal or when destination is not known	20	Peanuts to be subjected to sorting or other physical treatment, before human consumption, or use as an ingredient in food	15 (8)
Codex		Nuts and dried fruit to be subjected to sorting or other physical treatment, before human consumption, or use as an ingredient in food	10 (5)
Peanuts intended for further processing	15	Cereals and processed products thereof intended for direct human consumption or as an ingredient in food	4 (2)

[1]Food and Drug Administration (FDA) standard unless otherwise noted.
[2]Numbers in parentheses refer to separate standards for Aflatoxin B1 alone.

Source: Dohlman, 2003[17]

Veterinary Drugs and Antibiotic Residues

The veterinary drugs and antibiotic residues found in food products that are hazardous are antibiotics, growth-promoters, and anti-parasitic agents. These are used in the intensive farming of beef, sheep, pig, poultry, and fish. Use of these drugs is for prophylaxis (preventive care for those at risk), metaphylaxis (to improve disease resistance in clinically healthy animals) and therapeutic (treatment) purposes. Growth-promoters are used to boost the weight of the animals by increasing the efficacy of digestion and absorption and are fed to the animal in sub-therapeutic doses for extended periods of time[18]. Antiparasitic agents or Anthelmintics are used to treat parasites like warble-fly, fluke, and worms[19]. Examples of such drugs are Ivermectin, Benzimidazoles and Organophosphorus compounds.

With appropriate usage, drug residues should not remain in the resulting food products since the treatment is not recommended at the time approaching slaughter. It is the inadequate withdrawal period before slaughtering or milking that leads to the presence of residues in the meat/milk/eggs etc. Antibiotics could also be found in foods if they are used to extend the shelf life of meat and milk. The deleterious effects to humans consuming foods with drug residues are allergic reactions, disparities of inherent microflora and bacterial resistance to antibiotic treatments. There will also be negative effects to the production of fermented foods such as yogurt and cheese. Maximum permissible limits of certain antibiotics in meat in the USA are given in Table 3.

Table 3: Maximum permissible level of antibiotics in meat in the USA

Antibiotics	Animal Species	Level (mg/Kg)
Penicillin	Cattle, calves	0.05
Tetracycline	Calves, sheep, goats, pigs	0.25
Tylosine	Cattle, calves, pigs	0.20
Erythromycine	Pigs	0.10
Neomycine	Calves	0.25
Oxytetracycline	Cattle, calves, pigs	0.10
Chlortetracycline	Cattle	0.10
Chlortetracycline	Calves and pig muscle	1.00
Chlortetracycline	Calf liver and kidney	4.00
Chlortetracycline	Pig Liver	2.00
Lincomycine	Pigs	0.10

Source: Singh *et al.*, 2014[18]

Hormones

Hormones are used in animal agriculture to promote their growth and thus increase the muscle mass or increase milk production in case of female bovines. Risk assessment of these hormones consists of determining their microbiological and toxicological effects and identifying Acceptable Daily Intakes (ADI). Microbiological impact is usually on the gut microflora. Toxicological effect includes any deleterious effects caused to immunity, body weight and functions. A comparison of toxicological and microbiological risk assessment is given in Table 4.

Table 4: Comparison of toxicological and microbiological risk assessments

Item and Procedure	**Toxicological risk assessment**	**Microbiological risk assessment**
Target	Human, extrapolated from animals	Human Intestinal flora
Evaluation	General oral toxicity	Emergence of antimicrobial resistance in human intestinal normal microflora
	Genotoxicity	Disruption of barrier composed of human intestinal normal microflora
	Carcinogenicity	Change of metabolic microbiological activity
	Reproductive toxicity	
	Developmental toxicity	
	Immunotoxicity	
	Others	
	Observations in humans	
Endpoint	No-observed-adverse-effect-level (NOAEL) of the most sensitive toxic effect	No-observed-adverse-effect-concentration (NOAEC) of the most sensitive human gut flora
Uncertainty factor	Inter-species differences: 10	Data quality: 2-10
	Intra-species differences: 10	Type of most sensitive human flora
	Nature of toxic effects, data quality: 2-10	
Calculation of Acceptable Daily Intake (ADI)	NOAEL (μg/kgbw/day)/ uncertainty factor	(Min Inhibitory Conc. 50(μg/g) x Mass of colonic content (g)) / (Fraction of bio-available dose x Uncertainty factor x Body weight (kg))

Source: Jeong *et al.*, 2010[20]

Hormones in Meat

The natural hormones used in meat industries are progesterone, estradiol-17β and testosterone. Synthetic compounds used in this regard are trenbolone acetate with affinity to androgen receptors, zeranol with affinity for estrogen receptors and melengestrol acetate which acts like progestins[20]. These hormones are regarded as anabolic agents. These hormones are introduced into the animal body as ear implants and work towards increasing protein mass, decreasing fat, and improving feed efficiency and weight gain. Apart from this, beta-adrenergic agonists are added to the feed to increase the muscle to fat ratio. Optaflex, Paylean and Clenbuterol are commonly used beta-agonists[21].

Toxicological endpoints and regulatory limits of hormonal growth promoters are given in Table 5.

Table 5: Toxicological endpoints and regulatory limits of hormonal growth promoters

Compound	Toxicological endpoint	NOAEL (μg/kgbw/day)	ADI (μg/kgbw/day)	MRLs (μg/kg)
17β-estradiol	No relief of symptoms of menopause and changes in the serum concentrations of corticosteroid-binding globulin	5	0-0.05	Not necessary
Testosterone	Androgenic effects	1700	0-2	Not necessary
Progesterone	Changes in human uterus	3, 300(LOAEL)	0-30	Not necessary
Zeranol	Estrogenic effects	50	0-0-5	2(muscle), 10 (liver)
Melengestrol acetate	Changed menstrual cycle	5	0-0.03	1 (muscle), 10 (liver) 2(kidney), 18(fat)
Trenbolone acetate	Androgenic effects	2	0-0.02	2 (muscle, β-trenbolone) 10 (liver, α-trenbolone)

Source: Jeong *et al.*, 2010[20]

Hormones in Milk

The major steroid hormone residues found in milk are of glucocorticoids and adrenocorticotropic (ACTH) hormone. The normal concentration of glucocorticoid is 0.46 to 0.65 ng/ml [21]. These hormones help in the improvement of milk production. The commonly used synthetic hormones are dexamethasone (synthetic glucocorticoid) and ACTH. Use of ACTH leads to several fold increase in the milk cortisol levels. Melengestrol acetate (MGA) is progesterone used to suppress oestrus and improve weight gain in beef heifers. This compound has been found to be carcinogenic in several studies on rats and felids. In human females the Acceptable daily Intake (ADI) is 0.03 μg/kg/day[21]. Main estrogen in cow's milk is the biologically inactive 17β-estradiol, followed by estrone and 17α-oestradiol[21]. Insulin like Growth Factor 1 (IGF-1) is a polypeptide produced in liver and mammary cells of the bovine. IGF-1 production is increased when Bovine Somatotropic hormone is used on the animal. Increased levels of IGF-1 in humans have been associated with various types of cancer. Recombinant bovine growth hormone (rBGH) is a synthetic hormone used to increase production of milk in cows. rBGH is approved for use in USA, but not in European Union and Canada. The increased levels of rBGH leads to increase in IGF-1 as described above, which is carcinogenic in human body. Oxytocin is another protein hormone that helps in milk release and expulsion of placenta after delivery. Excessive amounts of this hormone in milk may lead to deleterious effects on humans such as nausea, blindness, and deafness in children.

Additives

The major additives used in the manufacture of food are coloring agents, enzymes, emulsifiers, stabilizers, flavoring agents, fortificants such as vitamins and minerals, pH control agents, antioxidants, antimicrobial agents, flavor enhancers, sweeteners, and polysaccharides. There are limits set to most of these additives by the Food Safety Standards Authority of India (FSSAI) and other regulatory agencies in various countries (Food and Drug Administration in USA). The use of food additives has enhanced the production of multiple varieties of shelf stable foods. While there are benefits, there are also concerns with respect to short- and long-term risks associated with consuming the additives. Food additives have made it possible to develop junk foods which have low available nutrients. Few examples of food additives that can cause hazardous reactions in people are described below:

i) Sulfites and sulfur dioxides are used in fruit products and wines. Sulfites cause asthma in persons who are allergic to them.

ii) Aspartame is an amino acid based low calorie sweetener used in soft drinks. This compound is hazardous to people with the genetic disease called Phenylketonuria since they cannot metabolize the phenylalanine which is a metabolic product of aspartame. Even high protein foods are of concern to such people.

iii) Nitrate and nitrites are additives used in cured meats for shelf-life extension and color preservation. Consumption of a large amount of this may cause Methemoglobinemia. This disease is characterized by bluish color of the skin and difficulty to breathe.

Biological Contaminants

Food contamination may originate from various biological sources. Bacteriological contaminants include harmful biological substances produced by different harmful microorganism viz., bacteria, virus, and fungi. The harmful pathogenic microorganisms contribute to the development of toxins which enters food and cause adverse health effects. The biological contaminants can enter food during preparation such as at the production stage, packaging stage, storage and transportation stage. The biological contaminants cause defects in the sensory characteristics/parameters such as texture, odour, taste, and nutritional quality of the food material[22].

Bacteria of Concern (general characteristics, different toxins)

Harmful pathogenic microorganisms viz., bacteria, fungi, virus, and parasites are prime culprits for biological contamination. The typical pathogens involved in food contamination are *Staphylococcus aureus, Listeria monocytogens, Salmonella enterica, Escherichia Coli O157:H7* and *Clostridium perfringens*. Serious food borne illnesses range from acute gastro enteritis to gastro intestinal cancers caused by food infection and food intoxication[23].

Coliforms and *E. coli*

E. coli is commonly found in intestine of the human beings and animals. However, majority of the strains are innoxious. Serious food contamination occurs by diverse group of serotypes. Entero-hemorrhagic *E. coli* (0157:H7) produces Shiga toxin and this shall be detected in raw foods viz., raw milk and undercooked ground beef[24].

Listeria monocytogenes

Listeria species are rod shaped bacteria, small, gram positive. They are facultative anaerobes i.e., can grow in anaerobic and aerobic conditions. *Listeria monocytogenes* commonly found in meat, poultry, and seafood products. The main causative organism for "Listeriosis" in humans is *Listeria monocytogenes*. The bacterium can grow at refrigeration temperature (0 – 4°C) and able to survive in a wide range of pH and salt concentration[25].

Salmonella sp.

Salmonella bacteria typically exist in the intestine of animals and humans. *S. enteritidies* and *S. typhimurium* are main causative organisms for Human Salmonellosis[26]. *Salmonella* is an enteric pathogen that attacks the gut mucosa and release toxins. Contaminated dairy products, eggs and meat products are the common cause for Salmonellosis[27].

Staphylococcus aureus

Staphylococcal food poisoning outbreak is caused by *Staphylococcus aureus*. *Staphylococcus aureus* can survive a wide range of temperature, pH, and salt concentration[28]. Due to these characteristics, the bacterium can grow in several food products such as meat, egg, milk and dairy products, salads, bakery products etc. Usually 3–5 hours after ingestion of foods contaminated with *S.aureus,* related symptoms such as hyper salivation, nausea, vomiting and diarrhea This is due to the production of various enterotoxins such as SEA, SEB, SEC, SED, SEE, SEG, SEH, SEI, and SEJ by the bacterium[29].

Bacillus cereus

Bacillus cereus can cause two types of food borne diseases based on the toxins produced during different phases of their growth. Enterotoxins that are released during the log phase can cause diarrhoea within 8-24 hours after ingestion. On the other hand, emetic toxins produced during the stationary phase result in vomiting within 1-6 hours of ingestion[30]. *Bacillus cereus* contaminates vegetables, dairy products and rice[31].

Clostridium perfringens and *Clostridium botulinum*

Clostridium spp. belong to the family of Bacillaceae. *Clostridium* spp. are rod-shaped gram-positive spore forming obligatory anaerobes. These bacteria are commonly present in intestinal tract of animals and soil[32]. Two most common spp. which cause food borne illnesses are *Clostridium perfringens* and *C. botulinum.* Fresh and canned vegetables and fruits, milk products and meat products are the most affected foods that are contaminated by these bacteria[33]. Type A toxin producing *Clostridium perfringens* is mainly responsible for food borne illness during sporulation. Within the gut they produce enterotoxins which result in vomiting, diarrhea, nausea, and abdominal cramps and rarely cause death[34]. But neurotoxins produced by *Clostridium botulinum* are fatal poisons. They can also lead to flaccid muscle paralysis[35].

Vibrio cholerae

Vibrio cholerae are Gram-negative bacteria that are generally considered non-invasive, enterotoxigenic organisms involved in gastroenteritis of varying severity. Foods including vegetables, fruits, seafood, dairy products, poultry and meat products and others can become contaminated with *Vibrio* spp. through improper handling, undercooking, washing with unhygienic water and using untreated night soil[36,37,38].

Shigella sp.

Shigella spp. such as *S.dysentriae, S.sonnei, S.flexneri and S.boydii* are enteric group of bacteria. *Shigella* spp. is main causative organism for "Shigellosis" in humans. This bacterium is rod shaped, Gram-negative, non-spore forming and immotile. Contaminated food and poor sanitation are the major causative factor for Shigellosis[39]. The symptoms are fever, nausea, vomiting, tenesmus and cramping which are manifested within 12 hours to 3 days after ingestion of the organism. *Shigella* causes the complications by direct invasion of colonic mucosa or production of enterotoxins[40].

Yersinia enterocolitica

Yersinia enterocolitica is a gram-negative rod from the *Enterobacteriaceae* family. *Yersinia* (*Y.*) *enterocolitica* is a major foodborne pathogen causing "Yersiniosis". Contaminated pork and pork products are believed to be the most important source of infection[41]. Eating or tasting raw or medium done pork is a significant risk factor for pathogenic *Y. enterocolitica* infection. Foodborne outbreaks are also associated with the consumption of untreated water, contaminated milk (unpasteurized or inadequately pasteurized), tofu or bean sprouts[42].

Fungi of Concern

Food-borne illnesses could be caused due to fungi or their byproducts, such as poisoning by mushrooms or mycotoxins. Some fungi that contaminate or spoil food are known pathogens such as *Alternaria, Aspergillus, Candida, Fusarium* and *Mucormycetes*[43,44].

Yeasts and Moulds

Majority of yeasts and molds are obligate aerobes that require free oxygen for growth. Their acid/alkaline requirement for growth is quite broad, ranging from pH 2 to above pH 9. Their optimal growth temperature ranging from 10-35°C is also broad, with a few species capable of growth below or above this range. Moisture requirements of foodborne molds are relatively low; most species can grow at a water activity (a_w) of 0.85 or less, although yeasts generally require a higher water activity. Both yeasts and molds can cause varying degrees of deterioration and decomposition of foods. They invade crops such as nuts, grains, beans, and fruits in fields before harvesting and during storage. The contaminated food may be slightly blemished, severely blemished, or completely decomposed, with the actual growth manifested by rot spots of various sizes and colors, unsightly scabs, slime, white cottony mycelium or highly colored sporulating molds. Abnormal flavors and odors may also be produced. Contamination of foods by yeasts and molds can result in substantial economic losses to producer, processor, and consumer. Several foodborne molds may also be hazardous to human or animal health because of their ability to produce toxic metabolites known as mycotoxins. Most mycotoxins are stable compounds that are not destroyed during food processing or home cooking. Certain foodborne molds and yeasts may also elicit allergic reactions or may cause infections[45].

Aspergillus

Aspergillus species produce various life-threatening biotoxins such as aflatoxins (AFTs), ochratoxins (OTA), patulin (PAT), citrinin (CIT), aflatrem (AT), secalonic acids (SA), cyclopiazonic acid (CPA), terrein (TR), sterigmatocystin (ST) and gliotoxin (GT), fumonisins (FBs) trichothecenes.

Mycotoxins – Aflatoxins B1, B2, M1, M2, Ochratoxin, Patulin, Deoxynivalenol.

Mycotoxins are common contaminants in crops such as rice, maize, wheat, and other foods. The molecular weight of Mycotoxins is low. *Penicillium* and *Aspergillus* are frequently found on growing cereals such as corn, wheat, barley in the field[46]. Fungal species under the genera *Fusarium*, *Aspergillus* and *Penicillium* are the most important mycotoxin-producing species[47]. Aflatoxins (AFTs) and Ochratoxins are the most vicious among all the mycotoxins produced by over 300 fungal species[48]. *A. parasiticus, A. nomius* and *A. flavus* produce secondary metabolites known as Aflatoxins. Aflatoxins are polyketide, lethal, cancer-causing agents. They are heat labile and are also responsible for many health problems. Once this toxin gets absorbed, affected human beings encounter symptoms such as hemorrhage, acute liver damage, edema, and death. Aflatoxins are naturally found in foods such as maize, legumes, rice, spices, crude vegetable oil, cotton seeds, coco beans. Aflatoxin (AFT) was shown to be the leading cause of Turkey X illness in England in the 1960s. Aflatoxins are naturally found in foods such as maize, legumes, rice, spices, crude vegetable oil, cotton seeds, cocobeans[49,50,51,52,53,54].

AFTs are mutagenic, teratogenic, genotoxic, hepatotoxic, and carcinogenic. Aflatoxin B1, B2, G1, and G2 are the several types of AFTs. Both AFG2 and AFB2 are safe to consume up until the point at which they are metabolised into AFG1 and AFB1, respectively. In the milch animals that feed on grains contaminated with these toxins, metabolites like AFB1 & AFB2 are hydrolyzed into AFM1 & AFM2 respectively. AFM1, a by-product of AFB1 that is also known to be harmful and carcinogenic, functions as a hydroxylated AFB1 metabolite in both humans and animals. Long-term AFT consumption or high intake has resulted in the death of Mammals. Acute aflatoxicosis is characterised by fatigue, appetite loss, fatty liver, jaundice, and decreased milk production in dairy calves. The maximum AFM1 content in milk was set by the US Food and Drug Administration (USFDA) at 0.05 g/kg[55,56,57,58,59,60,61,62,63,64,65].

Ochratoxins are a type of mycotoxin produced by the fungi found in various feeds and foods as secondary metabolites. They fall under the Ochratoxin A, Ochratoxin B, and Ochratoxin C categories. Specific species of Aspergillus and Penicillium produce ochratoxin. They are found in corn, rice, wheat, infant foods, meat, wine, beer, cheese oilseeds, vinegar, and feedstuffs. [66].

Ochratoxin, which is classified as a group 2B carcinogen, was recognized by the International Agency for Research on Cancer in 1993 as a powerful kidney carcinogen. In humans, the toxin has also potential to cause hepatotoxic, teretoxic, neurotoxic, nephrotoxic and immunotoxicity to humans. Ochratoxin is the root cause of "Balkan Endemic Nephropathy," a chronic illness that leads to kidney failure. The toxic effects of Ochratoxin include liver damage, nephropathy, cell apoptosis, cell autophagy, calcium homeostasis, protein synthesis suppression, and oxidative stress[67].

A poisonous metabolite called patulin (PAT) is produced by around 30 genera, primarily belongs to *Paecilomyces*, *Byssochlamys*, and *Aspergillus penicillium*. Patulin usually detected in fruits. Due to its toxic nature, Patulin causes ulcer, inflammation and intestinal heamorrage. There are established maximum PAT concentrations for juices (50 g/L) and solid apple products (25 g/L)[68,69].

The most common mycotoxins are Deoxynivalenol and its acetylated forms, 3-acetyl-deoxynivalenol (3-AcDON), and 15-acetyldeoxynivalenol (15-AcDON). DON is a common component of many grains, including wheat, corn, barley, rye, oat, and safflower seeds. It is primarily concentrated in the bran and outer layers of these grains[70].

Virus of concern – Picornaviruses, Reoviruses, Parvoviruses, Papovaviruses, Adenoviruses

The genus Hepatovirus of the family Picornaviridae includes the icosahedron enveloped virus species known as the hepatitis A viral (HAV), which has an approximately 7.5 kb ssRNA (+) genome. HAV can infect soil, food crops, and natural watercourses by sewage discharge[71,72]. Food is regarded as one of the main sources of HAV transmission to humans[73,74,75,76,77,78,79,80].

Mammalian Orthoreovirus (Reovirus) was the first naturally occurring double-stranded RNA (dsRNA) virus to be isolated from humans. In the 1950s, Albert Sabin and colleagues identified a novel virus with heteromorphologicvirions in the stools of children experiencing gastrointestinal distress[81]. Reovirus infections are often moderate or asymptomatic and spread through the respiratory or gastrointestinal pathways. Reovirus has been linked in recent investigations as a potential catalyst for the loss of oral tolerance to dietary antigens, which can result in celiac disease (CeD)[82].

Human-pathogenic parvovirus namely Human parvovirus B19 (B19V), which causes arthropathies and Erythema Infectious (a childhood rash), chronic anemia, aplastic crisis, and fetal death[83].

Papovaviruses are tiny, unenveloped, icosahedral viruses with double-stranded, circular DNA within. The diameter of viral particles varies from 45 to 55 nm.

Six or seven genes make up the small amount of genetic material found in papovaviruses, and the DNA's molecular weight ranges from 3×10^6 to 5×10^6 daltons.

Human adenoviruses (HAdVs) are a major cause of infections which can be fatal for immunocompromised individuals. They are associated with a wide range of clinical illnesses including upper and lower respiratory diseases, conjunctivitis, cystitis, and gastroenteritis. All HAdVs are excreted in faeces for an extended period[84,85]. The use of HAdVs as a viral indicator for water quality[86,87,88,89,90,91]. In water eight viral types of HAdVs belonging to A, C and F species were found[92]. Enteric type 41 (HAdV41), which causes diarrhoea, is the most common HAdV type found in wastewater and surface water[90].

Other Health Hazards

Allergens: 8 allergens

The proteins in food that cause allergic reactions are known as allergens[93,94]. These proteins typically resist heat during cooking, stomach acid. Iodine can be added to the diet to lessen the harmful effect of intestine digestive enzymes[95,96]. The main foods causing allergic reactions are milk, cereals, fish, egg, crustacean, tree nuts, shellfish, peanuts, soyabean, sulphite in concentrations of 10mg/kg or more.

Casein, β-lactoglobulin and α-lactalbumin are the main allergens associated with cow milk allergy [97]. IgE and mast cell response are involved in wheat allergy. Serine protease inhibitors, glutelins, and prolamins, among other wheat proteins, are the cause of the allergy[98].

Tree nut and Peanut allergies occur in 0.4-1.3% of children and 0.5-1% of adults[99,100]. The major classes of tree nut allergens include 11S globulins (legumins), 7S globulins (vicilins), and 2S albumins, all of which are rated as food storage proteins[101,102].

Shellfish allergy lasts the entirety of a person's life and is sometimes accompanied by severe symptoms, including anaphylaxis, which can be fatal[103]. The main allergen in shellfish and seafood is tropomyosin, a heat-stable muscle protein, with highly similar proteins also present in regularly consumed crustaceans[104].

The Food Allergy Safety, Treatment, Education and Research (FASTER) Act amended section 201(qq)(1) of the Federal Food, Drug, and Cosmetic Act, added sesame to the list of foods and food groups that are defined as major food allergens

Anti-nutritional substances (protease inhibitors, phytates, oxalates, tannins, haemagglutinnins, lectins, goitrogens/ anti-thyroid substances)

Protein-based compounds known as protease inhibitors can inhibit the activity of proteolytic enzymes in both humans and animals' gastrointestinal tracts. Phytates are compounds that inhibit the absorption of calcium and iron. Oxalates promote the production of kidney stones and decrease Ca absorption. Astringent, bitter, and polyphenolic in nature, tannins can be divided into two categories: condensed tannins and hydrolyzable tannins. Inhibition of digestive enzymes, increased excretion of endogenous protein, digestive system malfunctions, and toxicity of absorbed tannin or its metabolites are some of the harmful consequences of tannins. Haemagglutinnins, are proteins or glycoproteins. It prevents the small intestine from absorbing the digested end products. Goitrogens, which are naturally occurring chemicals that can affect the ability of the Thyroid gland to function, have been discovered in legumes including soybean and peanut. Goitrogens, which are sulfur-containing glycosides found in soy and cruciferous vegetables, induce the thyroid gland to enlarge by preventing the thyroid gland from absorbing iodine. Iodine can be added to the diet to lessen this harmful effect[105].

Plant toxins (favism, gossypol, erucic acid, cucurbitacins, BOAA–lathyrism, solanine, sanguinarine, linamarin, cyanide precursors – lima bean, tapioca, hydrocyanic acid, hypericin, saffrole, agaric acid, mushroom poisoning); seafood and Animal Toxins (shellfish poisoning, ciguatera toxin, scombroid poisoning, histamine, pufferfish poisoning, hallucinogenic fish poisoning, oyster abalone and red whelk poisoning, Azaspiracid poisoning, Brevetoxin, etc.).

Favism: Favism is observed in people with deficiency in glucose-6-phosphate dehydrogenase, an enzyme that is vital for the health of red blood cells. The impact of this disease is mostly on the males rather than females. Its relative occurrence is high in Near and Middle East, South America, Saharan Africa and South- East Asia. When certain substances such as broad beans, high amounts of vitamin C, and quinine are ingested, the red blood cells burst and cause anemia.

Gossypol:_Gossypol is a polyphenolic substance that is found in roots, seeds, and stem of *Gossypium* sp. (cotton). It stops spermatogenesis at small doses and hence can act as a male contraceptive.

Erucic acid: Erucic acid is a monounsaturated omega-9 fatty acid found in mustard and rapeseed oil. Prolonged consumption of erucic acid cause a temporary heart condition called myocardial lipidosis. European Food Safety Authority (EFSA) has set a tolerable daily intake of 7 milligrams per kilogram of body weight (mg/kg bw) per day.

Cucurbitacin: Cucurbitacin are found in the gourd and pumpkin family. In high concentrations they are toxic in nature and the resulting toxicity is referred to as "toxic squash syndrome".

Lathyrism: Lathyrism is a paralytic disease caused by prolonged feeding on species of Lathyrus such as the chickling pea (*L. sativus*), Lathyrus or the grass pea, *L. clymenum* and *L. cicera.*

Solanine: Solanine is a poisonous substance found in the nightshade family within the genus *Solanum* (particularly potatoes). It is a glycoalkaloid and if ingested causes neurological and gastrointestinal disorders.

Sanguinarine:Sanguinarineisfoundinbloodrootplants(Sanguinariacanadensis). Consumption of plants containing this compound results in Epidemic dropsy and sometimes is fatal.

Linamarin: Linamarin is found in the leaves and roots of flax, lima beans, and cassava. It is a cyanogenic compound. Linamarin produces cyanide when exposed to linamarase enzyme. Food preparations from cassava include blanching, boiling and fermentation which minimize and expel the cyanide derivatives which are volatile.

Safrole: Safrole is an organic compound found in Sassafras plants. It has a characteristic sweet-shop aroma and functions as an antifeedant protecting the plants from grazing animals and insects. All parts of Sassafras plants are used for culinary purposes. Studies in rats have shown that it causes cancer of liver at higher doses.

Agaric acid: Agaric acid is found in fungi. It has many medical uses. In lower forms of animals, it is found to suppress circulatory, respiratory, and nervous systems. It has been used as a metabolic inhibitor in animal research.

Mushroom poisoning: Mushroom poisoning is caused by the ingestion of mushrooms that produces various toxins. This happens when they are mistaken for edible mushrooms. Hence mushroom gatherers usually familiarize themselves with the outward characteristics of edible and non-edible mushrooms to prevent intoxication. Examples of mushroom toxins are alpha amanitin, phallotoxin, orellamine and muscarine etc.

Shellfish poisoning: Shellfish poisoning is caused by a toxin produced by certain algae and clams. Ingesting high amounts can cause paralysis and death. It is also known as Paralytic ShellFish poisoning.

Ciguatera: Ciguatera poisoning is caused by ingestion of fish which contains toxins such as maitotoxin and ciguatoxin. Symptoms include dizziness, vomiting and general weakness. It can last up to two days. *Gambierdiscus toxicus* is the marine organism that at first makes these toxins and is found along the reefs. Small fish ingest these which are again eaten by the larger fish. As the toxins move up the food chain, it gets more concentrated.

Pufferfish poisoning: Puffer fish or *Lagocephalus sceleratus* is a delicacy for many. But it contains tetrodoxin which can cause death by circulatory failure, and muscular paralysis. Hence it is a source of food poisoning with a high lethality rate.

Hallucinogenic fish poisoning: Icthyoallyeinotoxism is poisoning caused by consumption of certain fish. It causes alterations to the Central Nervous Systems resulting in hallucinations and nightmares.

Red Whelk poisoning: Red Whelk poisoning is caused by consumption of Neptuneaantiqua, a species of sea snail. It is characterized by visual disturbances and diarrhea.

Scombroid fish poisoning: Scombroid fish poisoning or histamine fish poisoning occurs after consuming fish high in histamine. Symptoms occur either within a few minutes or hours after ingesting the fish. The usual symptoms include facial flushing, dizziness, and rash etc. The irritation can last for several days or a few hours. This poisoning is caused when fish in the Scombridae family such as mackerel, tuna and skipjack are eaten after improper refrigeration. Bacteria break down the tissues generating histamines which are heat resistant.

Azaspiracid poisoning: Azaspiracids are toxins found in marine algae. They can accumulate in shellfish which cause poisoning when consumed by humans.

Brevetoxin poisoning: Brevetoxins are a set of chemicals produced by *Karenia brevis*. They are cyclic polyethers and cause illness known as neurotoxic shellfish poisoning.

Genetically Modified Microorganisms (GMOs)

Genetic engineering is a biological process that modifies the genetic code of all types of living things. According to WHO (World Health Organization), a GMO is an organism whose genetic material (DNA) has been altered in a way that does not occur normally through mating and/or natural recombination. These organisms include plants, animals, and microbes[106]. Toxicology, allergenicity, and genetic hazards are the three main health risks that could be linked to GM food. These may result from inserted genes and their expressed proteins, pleiotropic or secondary effects of gene-expression products, or even the disruption of native genes in the modified organism.

Some persons developed allergies to the beans produced by soybean plants that had a gene from Brazil nuts inserted into them due to the allergenicity that was detected in those plants. The GM maize line known as MON 863 is one of the most recent GM crops to be suspected of producing toxicity (YieldGard

Rootworm Corn). When the gene integrates into human DNA, they could switch on random genes inside of humans, resulting in an excess production of a toxin, allergy, or carcinogen[107].

Miscellaneous: Dioxins, acrylamide, radionuclides, Asbestos, VOCs, PHCs, PAHs, PCDDs, PCDFs, PCBs, PBDEs, PFASs, POPs, PPCPs, Plastics and synthetic polymers, MNMs, sulphites, melamine, etc.

Dioxins belong to the Persistent Organic Pollutants or the POP group. It is chemically known as 2,3,7,8- tetrachlorodibenzo para dioxin. Once they enter the body, they are highly stable and concentrate in the fat tissues. Dioxins are very toxic and cause damage to the immune system, cause cancer and developmental problems. Some of the most prominent Dioxins are as follows:

Acrylamide is a chemical produced in foods containing amino acid asparagine is heated to high temperatures in the presence of certain sugars, The International Agency for Research on Cancer describes acrylamide as "probable human carcinogen".

Radionuclide is an unstable radioisotope with excess nuclear energy and undergoes radioactive decay. Continued exposure to radionuclides is harmful to living beings and can cause cancer. Suitable radionuclides can also be used for treatment purposes in nuclear medicine.

Asbestos are fibrous materials resistant to corrosion. It is a commonly used building material for many products. If the fibers of asbestos are inhaled, they get lodged in the lungs and cause inflammation and scarring, subsequently causing breathing issues. Long exposures can cause cancer of chest and abdominal linings, ovary, lungs, and larynx.

VOCs are Volatile Organic Compounds that are volatilised from liquids or solids. They include an array of chemicals, with adverse health effects. Some examples of VOCs are methylene chloride, formaldehyde, ethylene glycol, and benzene etc. VOCs are released from fuels, solvents, paints, pesticides and personal care products.

Polycyclic aromatic hydrocarbons or PAHs_are common environmental pollutants. They are generated due to incomplete combustion of organic materials such as wood, petrol, coal, and oil. They have carcinogenic and toxic properties.

Polychlorinated dibenzodioxins (PCDDs) and dibenzofurans (PCDFs) are released from combustion of plastics and chlorinated waste. PCDDs and PCDFs are aromatic, planar, and tricyclic organic compounds.

PCBs or PolyChlorinated Biphenyls are organic pollutants. They are man-made compounds widely used in heat transfer fluids, lubricants, hydraulic fluids, and plasticizers.

PBDEs are Polybrominated diphenyl ethers which are synthetic organic compounds used in many household and commercial items to increase fire resistance. Studies in mice show that they cause many types of toxicities leading to cancer.

Perfluoroalkyl substances or PFASs are a large group of synthetic chemicals used in electronics, construction, automotive and military. They have a wide range of purposes. Exposure to PFAS can cause high cholesterol, hormone suppression, obesity, and cancer.

Pharmaceutical and personal care products or PPCPs are emerging contaminants. They leach into the agricultural soil systems from compost, wastewater and manure application. It accumulates in the plants and aquatic organisms and also causes toxicity.

Adulterants

Food Safety and Standards Act (2006) defined "adulterant" as any material which is or could be employed for making the food unsafe or sub-standard or mis-branded or containing extraneous matter. Food can be deemed adulterated, in accordance with the Federal Food, Drug, and Cosmetic Act (FFDCA), the main food safety regulation overseen by the Food and Drug Administration (FDA), if:

a) Cheaper or inferior quality item added to food
b) A substance is added which is injurious to health
c) Quality of food is below the standards
d) Any valuable constituent is extracted from main food article
e) To make it appear more valuable
f) Any substance has been added to increase bulk or weight

Synthetic Banned Colors

Non-permitted colours namely Metanil Yellow, Rhodamine B, Auramine, Malachite Green, Sudan dyes and Quinoline Yellow were identified in various foodstuffs. Studies have reported that erythrosine, tartrazine, amaranth, sunset yellow, carmoisine, picric acid, ponceau S, methyl orange and sudan dyes have been used as adulterants in saffron[108,109,110]. Commonly used color adulterants are given in Table 6.

Table 6: Commonly used color adulterants found in various foods.

S.No	Name of the adulterant	Food
1	Rhodamine B	Sweet potato, Ragi
2	Metanil Yellow	Turmeric
3	Anillin Dyes,	Turmeric
4	Lead (II) Chromate	Pulses
5	Malachite Green	Green pea and other green vegetable
6	Coal tar dyes	Oil
7	Sudan dyes	Chilli powder, Sweets and confectioneries
8	Auramine,	Sweets and confectioneries
9	Sudan dyes	
10	Quinoline Yellow	

Preservatives - formalin

Formaldehyde is also known as methanal and its chemical formula is HCHO. Formaldehyde is a colourless gas that has a strong, unpleasant odour when it is at room temperature[111]. Formalin is a generic term which describes a solution of 37 to 40 per cent formaldehyde gas dissolved in water[112]. Recently, International Agency for Research on Cancer (IARC) has classified formaldehyde as a Group 1 carcinogenic to humans[113].

To keep the freshness of fish and seafood, fishermen and fish vendors often tend to use formaldehyde solution[113] Calcium carbide, ethephon, formalin, colors and sweeteners are commonly used adulterants in Fruit. Harmful colours and Formalin are conventional adulterants in vegetables[120]. Sorbitol, Formalin, powdered milk or wheat flour, and detergent are used to adulterate milk [121]. Meat is frequently adulterated with toxic red color and formalin[122]. Sweetmeats are adulterated with toxic sodium cyclamate, colors and formalin [114].

Food fraud– filler materials, dilution, etc.

Food replacement, addition, tampering, misrepresenting food, food ingredients, or food packaging, as well as making false or deceptive claims about a product, are all collectively referred to as food fraud (Table 7). A more specific sort of fraud is the removal or substitution of genuine ingredients without the knowledge of the buyer to increase the seller's financial advantage[115]. Top ten products that are most at risk of food fraud are Fish, Olive oil, Milk, Honey & Maple syrup, Organic foods, Coffee and tea, Grains, Wines, Certain food juices and Spices[116].

Table 7: Different types of food fraud

S.No	Types	Example
1	Adulteration /dilution/ replacement	Olive oil diluted with potentially risk tree tea oil
2	Counterfeiting/Concealment	Poultry injected with hormones to conceal diseases
3	Artificial enhancement	Melamine added to enhance protein value
4	Use of non-declared, unapproved, or banned biocides	Use of unapproved antibiotic for animal for weight gain
5	Misrepresentation of nutritional content	Selling sunflower oil as cholesterol free
6	Fraudulent labeling claims	sale of inorganic foods as organic, selling non-halal meat products as halal
7	Removal of authentic constituents	removal of flavor components in cloves

Cross-Contaminations – Do's and Don'ts

Cross-contamination is the physical movement or transfer of harmful bacteria from one person, object or place to another. Preventing cross-contamination is a key factor in preventing foodborne illness

S.No	Do's	Don't s
1	Do Clean and disinfect work surfaces and all preparation tools thoroughly before starting food preparation	Don't Use same chopping boards and knives for raw meat/poultry and ready-to-eat food
2	Do Wash your hands before preparing food	Don't Store and keep raw and ready-to-eat food together.
3	Do Cover prepared meals, both cooked and otherwise.	Don't Use same cleaning materials, including cloths, sponges, and mops for raw and cooked food
4	Do Stack eggs, fish, and raw poultry below the prepared foods.	Don't store food in opened tins
5	Do Place raw meat, poultry, and seafood in containers or sealed plastic bags to prevent their juices from dripping onto other foods	Don't reuse freezer bags, cling film, or foil.
6	Do follow personal cleanliness (clean clothes, trimmed nails & waterproof bandage etc.) and personal behavior (hand washing, no loose jewelry, no smoking, no spitting etc.).	Don't let pest control baits, sprays and other chemicals come into contact with food, packaging, equipment or surfaces, because they are likely to be poisonous to people

Management of Hazards

To achieve a safe food, it is necessary to apply the hazard critical control point, GMP AND GHP concept and risk analysis to the food chain and to enact legislation, wherever necessary. Training to food handlers and monitoring and surveillance activities helps to improve the food safety Consumer Education is also needed to alert consumers regarding the risks arises from food and how to manage them.

Hazard Limits Set by Regulators – FSSAI, USFDA, EU and Codex

The scientific advice for Codex work is provided by FAO and WHO expert committees and ad hoc expert consultations. The limits for chemical hazard (additives, contaminants& mycotoxins, veterinary drugs), and biological hazards set by Joint FAO/WHO Expert Committee on Food Additives (JECFA); Joint FAO/WHO Meeting on Pesticide Residues (JMPR); & Joint FAO/WHO Expert Meetings on Microbiological Risk Assessment (JEMRA) which performs risk assessments. It lays down the maximum limits for certain food contaminants to protect the health of European Union (EU) citizens, including the most sensitive population groups, such as children, older people, and pregnant women. (Regulation (EC) No 1881/2006 setting maximum levels for certain contaminants in foodstuffs).

Title 21, Code of Federal Regulations, Part 110.110 allows the Food and Drug Administration (FDA) to establish maximum levels of natural or unavoidable defects in foods for human use that present no health hazard[117]. In India, Food Safety and Standards Authority of India has prescribed maximum limits for metal contaminants toxins and pesticide residues and antibiotic and other pharmacologically active substances in various food stuffs through Food Safety and Standards (Contaminants, Toxins And Residues) Regulations, 2011 [118].

HACCP, TACCP, VACCP, USFDA - Preventive Controls through GAP, GMP and GHP

HACCP (Hazard Analysis Critical Control Points):

To establish safety of the food meant for astronauts Hazard Analysis and Critical Control Point originally developed in the year 1960 by NASA. The 7 principles of HACCP is the base for food safety plans which called for determining what the potential hazards are for a business, how to identify if a hazard is exceeding acceptable levels, and what to do in such situations[119].

VACCP (Vulnerability Assessment Critical Control Points):

The VACCP system emphasizes on how to identify and control hazards. It also makes use of principles to HACCP. VACCP concentrate on food fraud principally concerned with economically motivated adulteration. VACCP is a systematic prevention of potential adulteration in a food, whether intentional or not, by identifying the vulnerable points in the supply chain[119].

TACCP (Threat Assessment Critical Control Points)

PAS 96:2017 defined TACCP as, "Systematic management of risk through the evaluation of threats, identification of vulnerabilities, and application of controls throughout the entire production process". Threats controlled by TACCP include intentional contamination, extortion, espionage, counterfeiting, and economically motivated adulteration (EMA).The goal of TACCP is to identify particular threats of criminal and malicious activity that could jeopardie food safety It requires users to put together a TACCP team, define the scope of the study, review current TACCP measures, characterize threats, develop mitigation strategies, scan the horizon for new or emerging threats, implement, record, and document, as well as audit / review, in accordance with the principles of Hazard Analysis and Critical Control Point (HACCP)[119].

HARPC (Hazard Analysis & Risk-Based Preventive Controls)

Hazard Analysis and Risk-Based Preventive Controls is a United States standard that is requisite under the Food Safety Modernization Act, 2011. With a focus on preventive control, it is an addition to the HACCP system. In other words, in place of HACCP plan which is used to identify hazards and control them when they occur, a HARPC plan focused to prevent the hazards occurring in the first place. This includes hazards including vulnerabilities and threats[119].

Label requirement /declaration of contaminants

Allergen declaration

As Per Food Safety and Standards (Labelling and Display) Regulations, 2020, The Foods And Ingredients Such As Cereals Containing Gluten, Crustacean And Their Products Milk & Milk Products, Eggs And Egg Products, Fish And Fish Products, Peanuts, Tree Nuts, Soybeans And Their Products, Sulphite In Concentrations Of 10mg/Kg Or More Which Are Known To Cause Allergy Shall Be Declared Separately in the product label as

Contains........................... (Name of allergy causing ingredients)

EU legislation has adopted precautionary labelling declaration allergens that enter accidentally in the food (cross-contamination) such as

"May Contain"- May contain Nuts"

To protect those with food allergies and other food hypersensitivities, the FDA enforces regulations requiring companies to list ingredients on packaged foods and beverages. For certain foods or substances that cause allergies or other hypersensitivity reactions, there are more specific labeling requirements. Either One of two ways must be used to identify the food source of the allergen on the food label.

The name of the food source of a major food allergen must appear:

In parentheses succeeding the name of the ingredient.

Examples: “flour (wheat),” “whey (milk)”and “lecithin (soy)”.

— OR —

Immediately close to or after the list of ingredients in a “contains” statement. Example: “Contains milk, soy, wheat.”

Conclusion

Food hazards are major threats to the society and its harmful effects impact our day-to-day life of everyone. To improve the foods safety among the consumers awareness training regarding the risk factors associated with food hazards should be conducted periodically through social media. Legal actions and penalty enforcement must be implemented to prevent introduction of food hazards in food supply chain such as processing, production, packaging, and transportation. Research studies shall be encouraged to adopt for less usage of pesticides, veterinary drugs, prolong and safe storage of grains and foods.

References

1.Adeyeye, S. A. (2016). Fungal mycotoxins in foods: A review. Cogent Food & Agriculture, 2, 1213127. [DOI: https://doi.org/10.1080/23311932.2016.1213127]

2.Andrade, P. D., & Caldas, E. D. (2015). Aflatoxins in cereals: worldwide occurrence and dietary risk assessment. World Mycotoxin Journal, 8(4), 415-431. [DOI: https://doi.org/10.3920/WMJ2014.1847].

3.Anza (2017). Food fraud fundamentals: Because you care about consumers' health. Merieuxnutrisciences

4.Arvanitoyannis, Ioannis S. [2008]. Comprehensive Analytical Chemistry: Food Contaminants and Residue Analysis Volume 51 || Chapter 2 International Regulations on Food Contaminants and Residues, pp. 29–76, [DOI:10.1016/S0166-526X (08)00002-0].

5.Asao, T., Buchi, G., Abdel-Kader, M. M., Chang, S. B., Wick, E. L., & Wogan, G. N. (1963). Aflatoxins b and g. Journal of the American Chemical Society, 85(11), 1706-1707. [DOI: https://doi.org/10.1021/ja00894a050]

6.Asi, M. R., Iqbal, S. Z., Ariño, A., & Hussain, A. (2012). Effect of seasonal variations and lactation times on aflatoxin M1 contamination in milk of different species from Punjab, Pakistan. Food control, 25(1), 34-38. [DOI: https://doi.org/10.1016/j.foodcont.2011.10.012]

7.Behar, A., Baker, K. S., Bassal, R., Ezernitchi, A., Valinsky, L., Thomson, N. R., & Cohen, D. (2018). Microevolution and patterns of transmission of Shigella sonnei within cyclic outbreaks shigellosis, Israel. Emerging infectious diseases, 24(7), 1335.

8.Belzile, Nelson; Chen, Yu-Wei; Filella, Montserrat (2011). Human Exposure to Antimony: I. Sources and Intake. Critical Reviews in Environmental Science and Technology, 41(14), 1309–1373. [DOI:10.1080/10643381003608227].

9.Bennett, C. A & Klich, M. A., (2003). Phylogenetic and morphological analysis of Aspergillus ochraceoroseus. Mycologia, 95(6), 1252-1260. [DOI: https://doi.org/10.1080/15572536.2004.11833033]

10.Bofill-Mas S, Calgua B, Clemente-Casares P, La Rosa G, Iaconelli M, Muscillo M, Rutjes S, de RodaHusman AM, Grunert A, Gräber I, Verani M, Carducci A, Calvo M, Wyn-Jones P, Girones R (2010) Quantification of human adenoviruses in European recreational waters. Food Environ Virol 2:101–109.

11.Bordin, K., Sawada, M. M., Rodrigues, C. E. D. C., da Fonseca, C. R., & Oliveira, C. A. F. (2014). Incidence of aflatoxins in oil seeds and possible transfer to oil: a review. Food Engineering Reviews, 6(1), 20-28. [DOI: https://doi.org/10.1007/s12393-014-9076-9]

12.Bosch, A. (1998). Human enteric viruses in the water environment: a minireview. Int Microbiol, 1(3), 191-6.

13.Bottone, E. J. (2015). Yersinia enterocolitica: revisitation of an enduring human pathogen. Clinical Microbiology Newsletter, 37(1), 1-8. (ISSN 0196-4399).

14.Brookes, G., & Barfoot, P. (2014). Economic impact of GM crops: the global income and production effects 1996–2012. GM crops & food, 5(1), 65-75. [DOI: https://doi.org/10.4161/gmcr.28098]

15.Brown, A. C., Cimmino, V. M., Szocik, J. F., Pass, H. A., Moline, S., De, S. K., & Domino, E. F. (2001). Allergic reactions to isosulfan blue during sentinel node biopsy—a common event. Surgery, 130(3), 439-442. [DOI: https://doi.org/10.1067/msy.2001.116407]

16.Clark, S., & Camargo Jr, C. A. (2005). Emergency management of food allergy: systems perspective. Current opinion in allergy and clinical immunology, 5(3), 293-298. [DOI: 10.1097/01.all.0000168797.14487.73]

17.Cook N & Rzeżutka A (2006) Hepatitis viruses. Emerging Foodborne Pathogens (Motarjemi Y & Adams M, eds), pp. 282–308. Woodhead Publishing Limited, Cambridge. (ISBN-10:1-84569-139-3).

18.Cullen, J. M., Ruebner, B. H., Hsieh, L. S., Hyde, D. M., & Hsieh, D. P. (1987). Carcinogenicity of dietary aflatoxin M1 in male Fischer rats compared to aflatoxin B1. Cancer research, 47(7), 1913-1917.

19.Dentinger CM Bower WA Nainan OV Cotter SM Myers G Dubusky LM Fowler S Salehi ED & Bell BP (2001) An outbreak of hepatitis A associated with green onions. J Infect Dis:183: 1273–1276. [DOI: https://doi.org/10.1086/319688].

20.Dermody TS, Parker JSL, Sherry B. (2013) Orthoreovirus. In: Knipe DM, Howley PM, editors. Fields virology. 6th ed. Philadelphia: Lippincott, Williams, & Wilkins; PP. 1304–46.

21.Di Stefano, Vita; Avellone, Giuseppe (2014). Food Contaminants. Journal of Food Studies, 3(1), 88–. doi:10.5296/jfs. v3i1.6192.[DOI:10.5296/jfs. v3i1.6192]

22.Divizia M Gabrieli R Donia D *et al.* (2004) Waterborne gastroenteritis outbreak in Albania. Water Sci Technol: 50: 57–61. [DOI: https://doi.org/10.2166/wst.2004.0018].

23.Dohlman, E. (2003). Mycotoxin hazards and regulations: Impacts on food and animal feed crop trade, international trade, and food safety: Economic theory and case studies, Jean Buzby (ed.), Agricultural Economic Report 828. USDA.

24.Eivazzadeh-Keihan, R., Pashazadeh, P., Hejazi, M., de la Guardia, M., & Mokhtarzadeh, A. (2017). Recent advances in nanomaterial-mediated bio and immune sensors for detection of aflatoxin in food products. TRAC Trends in Analytical Chemistry, 87, 112–128. [DOI: https://doi.org/10.1016/j.trac.2016.12.003]

25.Eivazzadeh-Keihan, R., Pashazadeh-Panahi, P., Baradaran, B., Miguel de la Guardia, Maryam Hejazi, Hessamaddin Sohrabi, Ahad Mokhtarzadeh, and Ali Maleki. (2018). Recent progress in optical and electrochemical biosensors for sensing of Clostridium botulinum neurotoxin. TRAC Trends in Analytical Chemistry, 103, 184–197, ISSN 0165-9936.

26.Ekramirad, N., Rady, A., Adedeji, A.A. and Alimardani, R. (2017) Application of Hyperspectral Imaging and Acoustic Emission Techniques for Apple Quality Prediction. Trans. ASABE, 60: 1391–1401. [DOI: 10.13031/trans.12184).

27.Elzupir, A. O., Alamer, A. S., & Dutton, M. F. (2015). The occurrence of aflatoxin in rice worldwide: a review. Toxin Reviews, 34(1), 37-42. [DOI: https://doi.org/10.3109/15569543.2014.984229]

28.Erdoğan, A., Ghimire, D., Gürses, M., Çetin, B., & Baran, A. (2018). Patulin contamination in fruit juices and its control measures. Avrupa Bilim ve Teknoloji Dergisi, (14), 39-48. [DOI: https://doi.org/10.31590/ejosat.434750]

29.Eyvazi, S., Baradaran, B., Mokhtarzadeh, A. and de la Guardia, M., 2021. Recent advances on development of portable biosensors for monitoring of biological contaminants in foods. Trends in Food Science & Technology, 114, pp.712-721. ISSN 0924-2244.

30.Farkas K, Cooper DM, McDonald JE, Malham SK, de Rougemont A, Jones DL (2018) Seasonal and spatial dynamics of enteric viruses in wastewater and in riverine and estuarine receiving waters. Sci Total Environ 634:1174–1183. [DOI: https://doi.org/10.1016/j.scitotenv.2018.04.038].

31.Feachem, R. G., Miller, C., & Drasar, B. (1981). Environmental aspects of cholera epidemiology. I. A review of selected reports of endemic and epidemic situations during 1961-1980. Tropical diseases bulletin, 78(8), 675-698.

32.Fernández, L., Méndez, J., & Guijarro, J. A. (2007). Molecular virulence mechanisms of the fish pathogen Yersinia ruckeri. Veterinary microbiology, 125(1-2), 1-10.

33.Fisher, E.L., Otto, M. and Cheung, G.Y., 2018. Basis of virulence in enterotoxin-mediated staphylococcal food poisoning. Frontiers in microbiology, 9, p.436.

34.Food Safety and Standards (Contaminants, Toxins and Residues) Regulation, 2011: www.fssai.gov.in/upload/uploadfiles/files/Compendium_Contaminants_Regulations_28_01_2022.pdf

35.Frisvad, J. C. (2018). A critical review of producers of small lactone mycotoxins: Patulin, penicillic acid and moniliformin. World Mycotoxin Journal, 11(1), 73-100. [DOI: https://doi.org/10.3920/WMJ2017.2294].

36.Galvano, F, Galofaro, V, & Galvano, G.1996). Occurrence and stability of aflatoxin M1 in milk and milk products: a worldwide review. Journal of Food protection, 59(10), 1079-1090. [DOI: https://doi.org/10.4315/0362-028X-59.10.1079].

37.Genchi, Giuseppe; Carocci, Alessia; Lauria, Graziantonio; Sinicropi, Maria Stefania; Catalano, Alessia (2020). Nickel: Human Health and Environmental Toxicology. International Journal of Environmental Research and Public Health, 17(3), 679–. [DOI:10.3390/ijerph17030679].

38.Granum, P. E. (2017). Spotlight on Bacillus cereus and its food poisoning toxins. FEMS microbiology letters, 364(10).

39.Greening GE (2006) Human and animal viruses in food (including taxonomy of enteric viruses). Viruses in Foods (Goyal SM, ed.), pp. 5–42. Springer, New York, NY.

40.Griffiths, M. W., & Schraft, H. (2017). Bacillus cereus food poisoning. In Foodborne diseases (pp. 395-405). Academic Press. ISBN 9780123850072.

41.Guan, Q., Yang, H., Zhao, Y., Han, Z., Ling, Z., Yang, K., Yin, C., Yu, S. [2021]. Microplastics release from victuals packaging materials during daily usage. EcoMat, 3(3): 1-9 [DOI:10.1002/eom2.12107].

42.Hourihane, J. O. B., Berger, A., Roberts, S. A., & Warner, J. O. (1998). Resolution of peanut allergy: case-control study Science commentary: Why do some children grow out of peanut allergy? Bmj, 316(7140), 1271-1275. [DOI: https://doi.org/10.1136/bmj.316.7140.1271]

43.Hourihane, J.O.B., 2011. Peanut allergy. Pediatric Clinics, 58(2), pp.445-458.
44.Huq, A., Small, E. B., West, P. A., Huq, M. I., Rahman, R. and Colwell, R. R. (1983). Ecological relationships between Vibrio cholerad planktonic crustacean copepods. Applied and Environmental Microbiology 45: 275-283, ISSN:0041-3240.
45.Hutin YJF Pool V Cramer EH *et al.* (1999) A multistate, foodborne outbreak of hepatitis A. N Engl J Med340: 595–602. [DOI:10.1056/NEJM199902253400802].
46.Ingredients, Additives, GRAS & Packaging Guidance Documents & Regulatory Information: https://www.fda.gov/food/guidance-documents-regulatory-information-topic-food-and-dietary-supplements/ingredients-additives-gras-packaging-guidance-documents-regulatory-information)
47.Inomata, N. (2009). Wheat allergy. Current opinion in allergy and clinical immunology, 9(3), 238-243. [DOI 10.1097/ACI.0b013e32832aa5bc]
48.Jarvis, N. A., O'Bryan, C. A., Dawoud, T. M., Park, S. H., Kwon, Y. M., Crandall, P. G., *et al.* (2016). An overview of Salmonella thermal destruction during food processing and preparation. Food Control, 68, 280–290, (ISSN 0956-7135).
49.Jayaraj, R., Megha, P., & Sreedev, P. (2016). Organochlorine pesticides, their toxic effects on living organisms and their fate in the environment. Interdisciplinary toxicology, 9(3-4), 90-100. [DOI: https://doi.org/10.1515/intox-2016-0012].
50.Jeong, M, S., Kang, D., Lim, M, W., Kang, C, S., & Sung. H, J. (2010). Risk Assessment of Growth Hormones and Antimicrobial Residues in Meat. Toxicol. Res. Vol 26, No. 4. Pp 301-313.
51.Joshi, V., Pandit, A., & Vaidya, D. (2017). Ochratoxin A in food: An overview. International Journal of Food and Fermentation Technology, 7, 13–24. ISSN :2277-9396
52.K.M. Amanda, R.L. Stephen, (2019) Mechanisms underlying aflatoxin-associated mutagenesis–implications in carcinogenesis, DNA Repair 77, 76–86. [DOI: https://doi.org/10.1016/j.dnarep.2019.03.004]
53.Kaiser J. (2000). Endocrine disrupters: Panel cautiously confirms low-dose effects. Science 290: 695–697. [DOI: 10.1126/science.290.5492.695].
54.Kamala, Katepogu [2018]. Microbial Contamination and Food Degradation || Food Products and Food Contamination, pp.1–19, [DOI:10.1016/B978-0-12-811515-2.00001-9].
55.Kőszegi, T., & Poór, M. (2016). Ochratoxin A: molecular interactions, mechanisms of toxicity and prevention at the molecular level. Toxins, 8(4), 111. [DOI: https://doi.org/10.3390/toxins8040111]
56.Kowalska, Aleksandra; Soon, Jan Mei; Manning, Louise [2018]. A study on adulteration in cereals and bakery products from Poland including a review of definitions. Food Control, S0956713518302329–. [DOI: 10.1016/j.foodcont.2018.05.007].
57.La Rosa G, Pourshaban M, Iaconelli M, Muscillo M (2010) Quantitative real-time PCR of enteric viruses in influent and effluent samples from wastewater treatment plants in Italy. Ann Ist Super Sanita 46:266–273. [DOI: 10.4415/Ann_10_03_07].
58.Labbe, R. G., & Juneja, V. K. (2017). Clostridium perfringens. Foodborne diseases. Elsevier pp. 235–242
59.Lee, D., & Lee, K. G. (2015). Analysis of aflatoxin M1 and M2 in commercial dairy products using high-performance liquid chromatography with a fluorescence detector. Food Control, 50, 467-471. [DOI: https://doi.org/10.1016/j.foodcont.2014.09.020]
60.Lee, J.H., Noh, J., Noh, G., Kim, H.S., Mund, S.H., Choi, W.S., Cho, S. and Lee, S. (2010). Allergen-specific B cell subset responses in cow's milk allergy of late eczematous reactions in atopic dermatitis. Cellular Immunol., 262: 44–51. [DOI: https://doi.org/10.1016/j.cellimm.2009.12.006].
61.Lees D (2000) Viruses and bivalve shellfish. Int J Food Microbiol: 59: 81–116. [DOI: 10.1016/s0168-1605(00)00248-8].

62.Lien, K. W., Wang, X., Pan, M. H., & Ling, M. P. (2019). Assessing aflatoxin exposure risk from peanuts and peanut products imported to Taiwan. Toxins, 11(2), 80. [DOI: https://doi.org/10.3390/toxins11020080]

63.Lozano, P., Castellar, M. R., Simancas, M. J., &Iborra, J. L. (1999). Quantitative highperformance liquid chromatographic method to analyse commercial saffron (Crocus sativus L.) products. Journal of Chromatography A, 830, 477–483. [DOI: https://doi.org/10.1016/S0021-9673(98)00938-8]

64.Luoma, Samuel N.; Presser, Theresa S. (2009). Emerging Opportunities in Management of Selenium Contamination ¹. Environmental Science & Technology, 43(22), 8483–8487. [DOI:10.1021/es900828h].

65.M Müller; M Anke; H Illing-Günther (1998). Aluminium in foodstuffs., 61(4), 419–428. [DOI:10.1016/s0308-8146(97)00085-x].

66.MacLennan, C. A., Riddle, M. S., Chen, W. H., Talaat, K. R., Jain, V., Bourgeois, A. L., *et al.* (2019). Consensus report on Shigella controlled human infection model: Clinical endpoints. Clinical Infectious Diseases, 69, S591–S595.

67.Mahfuz, M. and Mahin, D. 2014. BahariIftarirShomahar, The Daily AmaderShomoy, 1 July 2014, http://www.dainikamadershomoy.com/

68.Martha IS Verissimo; João ABP Oliveira; M. Teresa SR Gomes (2005). The evaluation of copper contamination of food cooked in copper pans using a piezoelectric quartz crystal resonator., 111-112(none), 587-591. [DOI: 10.1016/j.snb.2005.05.004].

69.Mena, K. D., & Gerba, C. P. (2008). Waterborne adenovirus. Reviews of environmental contamination and toxicology, 133-167

70.Moore, G. G., Mack, B. M., Beltz, S. B., & Puel, O. (2018). Genome sequence of an aflatoxigenic pathogen of Argentinian peanut, Aspergillus arachidicola. BMC genomics, 19(1), 1-12. [DOI: https://doi.org/10.1186/s12864-018-4576-2.]

71.Nagaraj G. (2009), Oil Seeds- Properties, Processing, Products and Procedures. New India Publishing Agency, New Delhi, PP: 601. (ISBN- 978-81-90723-75-6)

72.Nerín, Cristina; Aznar, Margarita; Carrizo, Daniel [2016]. Food contamination during food process. Trends in Food Science & Technology, 48: 63–68. [DOI:10.1016/j.tifs.2015.12.004].

73.Ngindu, A., Kenya, P., Ocheng, D., Omondi, T., Ngare, W., Gatei, D., Johnson, B., Ngira, J., Nandwa, H., Jansen, A. and Kaviti, J., 1982. Outbreak of acute hepatitis caused by aflatoxin poisoning in Kenya. The Lancet, 319(8285), pp.1346-1348. [DOI: https://doi.org/10.1016/S0140-6736(82)92411-4]

74.Noordiana, N., Fatimah, A. B., & Farhana, Y. C. B. (2011). Formaldehyde content and quality characteristics of selected fish and seafood from wet markets. International Food Research Journal, 18(1).

75.Nygård K, Andersson Y & Lindkvist P *et al.* (2001) Imported rocket salad partly responsible for increased incidence of hepatitis A cases in Sweden, 2000–2001. Euro Surveil: 151–153. [DOI: https://doi.org/10.2807/esm.06.10.00380-en].

76.O'keeffe, M., & Kennedy, O. (1998). Residues – a Food Safety Problem? Journal of Food Safety, 18(4), 297–319. [DOI: https://doi.org/10.1111/j.1745-4565.1998.tb00222.x]

77.Ogorzaly L, Walczak C, Galloux M, Etienne S, Gassilloud B, Cauchie HM (2015) Human adenovirus diversity in water samples using a next-generation amplicon sequencing approach. Food Environ Virol 7:112–121.

78.Ostry, V., Malir, F., Toman, J., & Grosse, Y. (2017). Mycotoxins as human carcinogens—the IARC Monographs classification. Mycotoxin research, 33(1), 65-73. [DOI: https://doi.org/10.1007/s12550-016-0265-7]

79.Otero, V., Sanchez, S., Herrera-Leon, S., Rodríguez-Calleja, J.-M., Otero, A., GarcíaLopez, (2017). Detection and characterization of Shiga toxin-producing Escherichia coli (STEC) in bulk tank ewes' milk and sheep farm environment. Small Ruminant Research, 154, 110–114(ISSN:0921-4488).

80.Ozluoymak, O. B., & Guzel, E. (2018). Prediction of aflatoxin contamination on dried fig (ficus carica) samples by spectral image analysis in comparison with laboratory results. Fresenius Environmental Bulletin, 27, 681-689.

81.Patel, K. J., & Boghra, V. R. (2018). Appraisal of various physico-chemical characteristic and detection of synthetic color in saffron (Crocus sativus). Asian Journal of Dairy & Food Research, 37(4). Print ISSN:0971-4456 / Online [ISSN:0976-0563].

82.Pebody RG, Leino T, Ruutu P, Kinnunen L, Davidkin I, Nohynek H & Leinikki P (1998) Foodborne outbreaks of hepatitis A in a low endemic country: an emerging problem? Epidemiol Infect: 120: 55–59. [DOI: https://doi.org/10.1017/S0950268897008340]

83.Pernin, A., Guillier, L., & Dubois-Brissonnet, F. (2019). Inhibitory activity of phenolic acids against Listeria monocytogenes: Deciphering the mechanisms of action using three different models. Food Microbiology, 80, 18–24. (ISSN:0740-0020).

84.Petrakis, E. A., Cagliani, L. R., Polissiou, M. G., & Consonni, R. (2015). Evaluation of saffron (Crocus sativus L.) adulteration with plant adulterants by 1H NMR metabolite fingerprinting. Food Chemistry, 173, 890-896. [DOI: https://doi.org/10.1016/j.foodchem.2014.10.107]

85.Pires, P. N., Vargas, E. A., Gomes, M. B., Vieira, C. B. M., Santos, E. A. D., Bicalho, A. A. C., & Trovatti Uetanabaro, A. P. (2019). Aflatoxins and ochratoxin A: occurrence and contamination levels in cocoa beans from Brazil. Food Additives & Contaminants: Part A, 36(5), 815-824. [DOI: https://doi.org/10.1080/19440049.2019.1600749]

86.Pitt, J. I., & Hocking, A. D. (2009). Fungi and food spoilage (Vol. 519, p. 388). New York: Springer. (ISBN: 978-0-387-92207-2).

87.Qiu J, Söderlund-Venermo M, Young NS. 2017. Human parvoviruses. Clin. Microbiol. Rev. 30:43–113. [DOI: https://doi.org/10.1128/CMR.00040-16].

88.R.G. Liteplo, R.B., M.E. Meek, R. Chénier, FORMALDEHYDE. 2002, World Health Organization: Geneva.

89.Rabbani, G. H. and Greenough, W. B. (1999). Food as a Vehicle of Transmission of Cholera. Journal of Diarrhoeal Diseases Research 17 (1): 1-9.

90.Rahman, M. M., Ahmed, S., Hosen, M. M., & Talukder, A. K. (2012). Detection of formalin and quality characteristics of selected fish from wet markets at Sylhet city in Bangladesh. Bangladesh Research Publications Journal, 7(2), 161-169. [ISSN: 1998-2003].

91.Rames E, Roiko A, Stratton H, Macdonald J (2016) Technical aspects of using human adenovirus as a viral water quality indicator. Water Res 96:308–326. [DOI: https://doi.org/10.1016/j.watres.2016.03.042].

92.Rath, A. P., Panda, S., Pandey, R., Routray, A., & Sethy, K. (2018). Hormone residues in milk and meat products and their public health significance. 6. The Pharma Innovation Journal; 7(1): 489-494.

93.Rather, I. A., Koh, W. Y., Paek, W. K., & Lim, J. (2017). The sources of chemical contaminants in food and their health implications. Frontiers in pharmacology, 8, 830.

94.Renée Johnson (2014). Food Fraud and "Economically Motivated Adulteration" of Food and Food Ingredients. Congressional Research Service, 7-5700. www.CRS.gov.R43358

95.Rodríguez-Lázaro D, Cook N, Ruggeri FM, Sellwood J, Nasser A, Nascimento MS, D'Agostino M, Santos R, Saiz JC, Rzeżutka A, Bosch A, Gironés R, Carducci A, Muscillo M, Kovač K, DiezValcarce M, Vantarakis A, von Bonsdorff CH, de

RodaHusman AM, Hernández M, van der Poel WH (2012) Virus hazards from food, water and other contaminated environments. FEMS Microbiol Rev 36:786–814

96.Roux, K.H., Teuber, S.S. and Sathe, S.K. (2003). Tree nut allergens. Internat. Arch Allergy Immunol., 131: 234-244.

97.Rushing, B. R., & Selim, M. I. (2019). Aflatoxin B1: A review on metabolism, toxicity, occurrence in food, occupational exposure, and detoxification methods. Food and chemical toxicology, 124, 81-100. [DOI: https://doi.org/10.1016/j.fct.2018.11.047]

98.Rusiñol M, Fernandez-Cassi X, Timoneda N, Carratalà A, Abril JF, Silvera C, Figueras MJ, Gelati E, Rodó X, Kay D, Wyn-Jones P, Bofill-Mas S, Girones R (2015) Evidence of viral dissemination and seasonality in a Mediterranean river catchment: implications for water pollution management. J Environ Manage 159:58–67. [DOI: https://doi.org/10.1016/j.jenvman.2015.05.019].

99. S. Jacob, J. T. Fernandez, P. C. Gupta, M. Raman, J. A. John. (2022) Food hazards and their management systems. Food Processing and Preservation Technology 679-702 [ISBN: 978-9-39-106385-6]. Narendra Publishing House, Delhi, India

100.Sabin, A. B. (1959). Reoviruses: a new group of respiratory and enteric viruses formerly classified as ECHO type 10 is described. Science, 130(3386), 1387-1389. DOI: 10.1126/science.130.3386.138.

101.Sathe, S.K., Kshirsagar, H.H. and Roux, K.H. (2005). Advances in seed protein research: a perspective on seed allergens. J. Food Sci., 70: 93-120. [DOI: https://doi.org/10.1111/j.1365-2621.2005.tb11454.x]

102.Shrestha, A., Uzal, F. A., & McClane, B. A. (2018). Enterotoxic clostridia: Clostridium perfringens enteric diseases. Microbiology spectrum, 6(5), 6-5.

103.Sicherer, S.H. 2000. Determinants of systemic manifestations of food allergy. J. Allergy Clin. Immunol.,106: 251-257.

104.Singh, S., Shukla, S., Tandia, N., Kumar, N., & Paliwal, R. (2014). ANTIBIOTIC RESIDUES: A GLOBAL CHALLENGE. 14. Pharma Science Monitor 5(3), Jul-Sep 2014, 184-197.

105.Sinnhuber, R. O., Lee, D. J., Wales, J. H., Landers, M. K., & Keyl, A. C. (1974). Hepatic carcinogenesis of aflatoxin M1 in rainbow trout (Salmo gairdneri) and its enchancement by cyclopropene fatty acids. Journal of the National Cancer Institute, 53(5), 1285-1288. [DOI: https://doi.org/10.1093/jnci/53.5.1285]

106.Sohail E, Waseem A, Chae WL, Jong JL, Imitiaz H. (2004). Endocrine Disrupting Pesticides: A Leading Cause of Cancer among Rural People in Pakistan. Experimental Oncology 26(2): 98–105.

107.Steensma, D.P. (2003). The kiss of death: a severe allergic reaction to a shellfish induced by a good-night kiss. Mayo. Clin. Proc., 78: 221-222. [DOI: https://doi.org/10.4065/78.2.221]

108.Su Y, Dai Y, Lin Y, Gao X, Han Y, Zhao B. (2012). Serum organochlorine pesticide residues and risk of gallstone disease: a case-control study in Xiamen. Annals of Epidemiology 22(8): 592–597. [DOI: https://doi.org/10.1016/j.annepidem.2012.05.002]

109.Swainson, M. (2019). Food sector challenges and the role of technical and quality management. In, Swainson's Handbook of Technical and Quality Management for the Food Manufacturing Sector. Cambridge: Woodhead Publishing, pp. 1-606

110.Tallon LA, Love DC, Moore ZS & Sobsey MD (2008) Recovery and sequence analysis of hepatitis A virus from spring water implicated in an outbreak of acute viral hepatitis. Appl Environ Microbiol: 74: 6158–6160. [DOI: https://doi.org/10.1128/AEM.02872-07].

111.The Daily Star, 12 July 2007. Adulterated ghee seized. http:// www.thedailystar.net/newDesign/news-details.php?nid =66453.

112.The Daily Star, 30 January 2007. Traders fined for selling toxic chemicals as food colour. http://archive.thedailystar.net/ 2007/01/30/d70130013625.htm.

113.The daily star,1 november 2009.trader finded for selling fish treated with formalin (http:// archive.thedailystar.net/new Design/news-details.php?nid=103948).

114.Tomsikova, A. (2002). Risk of fungal infection from foods, particularly in immunocompromised patients. Epidemiologie, mikrobiologie, imunologie: casopis Spolecnosti pro epidemiologii a mikrobiologii Ceske lekarske spolecnosti JE Purkyne, 51(2), 78-81.

115.Winston RL, Schwarzla¨nder M, Hinz HL, Day MD, Cock MJW, Julien MH (2014) Biological control of weeds: a world catalogue of agents and their target weeds, 5th edn. FHTET-2014-04. Pg:838. USDA Forest Service, Forest Health Technology Enterprise Team, Morgantown.

116.World health organization, 2016 http://wwwwhoint/foodsafety/areas work/food-technology/faq-genetically-modified-food/en/.

117.World Health Organization, Food and Agriculture Organization of the United Nations & Joint FAO/WHO Expert Consultation on Allergenicity of Foods Derived from Biotechnology (2001: Rome, Italy). (2001). Evaluation of allergenicity of genetically modified foods: report of a Joint FAO/WHO Expert Consultation on Allergenicity of Foods Derived from Biotechnology, 22 – 25 January 2001. Food and Agriculture Organization of the United Nations. https://apps.who.int/iris/handle/10665/340572, PP. 22-25

118.Wu, W., & Zeng, L. (2017). Current and emerging innovations for detection of food-borne Salmonella. Current Topics in Salmonella and Salmonellosis, 83.

119.Wyn-Jones AP, Carducci A, Cook N, D'Agostino M, Divizia M, Fleischer J, Gantzer C, Gawler A, Girones R, Höller C, de RodaHusman AM, Kay D, Kozyra I, López-Pila J, Muscillo M, Nascimento MS, Papageorgiou G, Rutjes S, Sellwood J, Szewzyk R, Wyer M (2011) Surveillance of adenoviruses and noroviruses in European recreational waters. Water Res 45:1025–1038. [DOI: https://doi.org/10.1016/j.watres.2010.10.015].

120.Y.M. Abdulrazzaq, R. Padmanabhan, S. Bastaki, J. Kochyil, M. Shafiullah, (2011) Teratogenic effects of aflatoxin B1 in mice exposed in early and late gestation, Pediatr. Res. 70, 405–405. [DOI: https://doi.org/10.1038/pr.2011.630]

121.Yadrykhins'ky, V. S. (2018). Clostridium spp. and modeling the butanol producer on its basis. Science.

122.Yeasts, Molds and Mycotoxins: https://www.fda.gov/food/laboratory-methods-food/bam-chapter-18-yeasts-molds-and-mycotoxins.

6

A Review on Biopolymers and Its Application in Food and Allied Industry

Twinkle Kesharwani, Meenakshi Garg* and Susmita Dey Sadhu

Bhaskaracharya College of Applied Sciences, Dwarka, New Delhi

Abstract

The food packaging industry relies heavily on various materials like glass, paper, wood, and plastics. Plastics, in particular, are becoming increasingly valuable in both rigid and flexible packaging due to their advantages such as low weight, good formability, and versatility. However, despite their numerous benefits, plastics are primarily non-biodegradable and contribute to environmental pollution. With growing consumer awareness about health, hygiene, and the environment, coupled with increasing purchasing power, there's a rising demand for eco-friendlier options that are both affordable and sustainable. As a result, we are seeing a significant increase in the use of biopolymers as an alternative to plastics in food packaging. The term "bio" indicates that these plastics are derived from natural sources (bio-based). Their biodegradability and non-toxicity make them a sustainable alternative for food packaging.

This review article will discuss the types of biopolymers, their properties, applications in food and related industries, and their processing methods.

Keywords: *Biopolymer, Bioplastic, Packaging Material, Biodegradable, Food Industry*

Introduction

A biopolymer is defined as a polymeric substance obtained from any living organism (plant or animal) [1]. These polymers are organic compounds where simpler molecules are covalently bonded with each other [2]. Although biopolymers are strictly synthesized naturally by living cells but some of them are also synthesized from non-biomaterial as well. Natural biopolymers are of different types based on their chemical nature like carbohydrate based, protein based or lipid based. Biopolymers can be broken down by the action of naturally-occurring organisms which leaves organic non-harming by-products like water and carbon-dioxide in the environment[3].

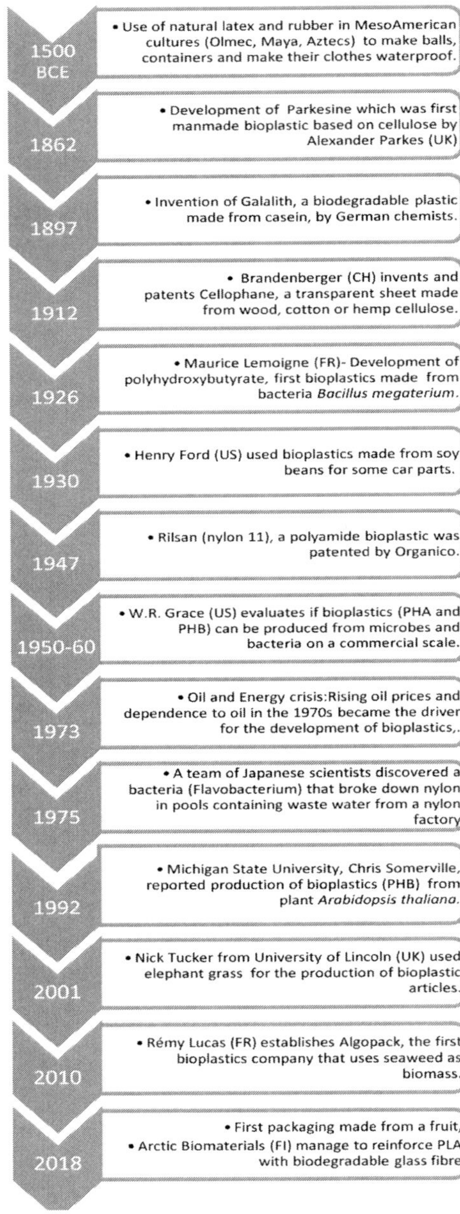

Bio polymers are gaining importance in the field of food packaging, medical application, commodity packaging etc. because of their have specific physical, chemical, biological, biomechanical and degradation properties. Wide ranges of natural or synthetic biopolymers capable of undergoing degradation hydrolytically or enzymatically are being investigated for many applications[1]. Bio polymers are mainly based on two properties: a) Biological origin b) Biodegradability.

Earlier natural biopolymers such as cellulose, protein, starch have been used extensively for food applications. Recently, synthetic biopolymers such as polylactic acid (PLA), polycaprolactone

(PCL), polyglycolic acid (PGA), polyvinyl alcohol (PVOH), and polybutylene succinate (PBS) are gaining more attention[4] for various applications. Natural biopolymers are more stereospecific and mostly hydrophilic having complex 3D structures as compared to the synthetic polymers[2]. Among the synthetic biopolymers some are microbiologically degradable in the environment whereas some are recyclable[1].

Biopolymers and their derivatives are gaining much interest because of their unique properties like biodegradability, availability, nontoxicity, and compatibility[5].

Throughout history, human used biopolymers as food, manufacture of clothing and furniture and packaging[6] but recently it has gained much interest with advancement in technologies. Figure-1 is showing history and important innovations in field of biopolymers.

Broadly biopolymers provide some advantages[7] such as

- Biodegradability
- Biocompatibility
- Produced from natural sources
- Can be recycled and reused
- Burnable and compostable without toxic byproducts
- Provide low carbon footprint

Global Status of Biopolymers

Global market for biopolymer is booming across the world especially in developed and developing nations like North America, Europe, China, Japan, Southeast Asia and India, it is expected to increase manifold in near future. According to the latest market data (Figure-2) compiled by European Bioplastics in cooperation with the research institute nova-Institute[8], currently, bioplastics represent about 1% of more than 359 million tonnes of plastic that is produced annually but their demand is rising and the market for biopolymers is growing and diversifying continuously. The global bioplastics production capability is set to increase from about 2.11 million tonnes in 2019 to around 2.43 million tonnes in 2024. The more growth has been seen in case of latest and innovative biopolymers like bio-based PP, PHA due to their diverse range of applications in various sectors. Their production capacities are predicted to be increase by an exponential rate by 2024. Bio-based, non-biodegradable plastics altogether including bio-based PE, bio-based PET and bio-based PA, currently make up for over 44% i.e. almost 1 million tones of the global bioplastics production capacities. Biodegradable plastics altogether, including, starch blends, PHA, PLA and others, account for over 55.5% of global bioplastics. The production of biodegradable polymer is expected to increase to 1,33 million in 2024 mainly due to significant growth rates of PHA.

In India biopolymer has a massive scope in various applications but because of many challenges, the Indian biopolymer market are still in their embryonic stage. In present very few commercial players are operating in this segment[7]. There are chances that recent ban on single use plastic will increase the market value of biopolymers in India. However promotion of long term benefits of biopolymers as well as environmental awareness are the important steps needed to be taken to overcome these challenges.

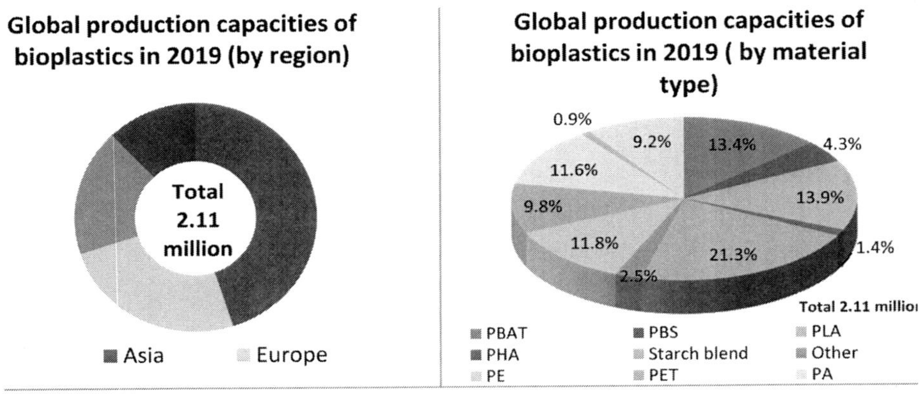

Figure 2: Global production capacities of bioplastics [8]

3. Classification of Biopolymers

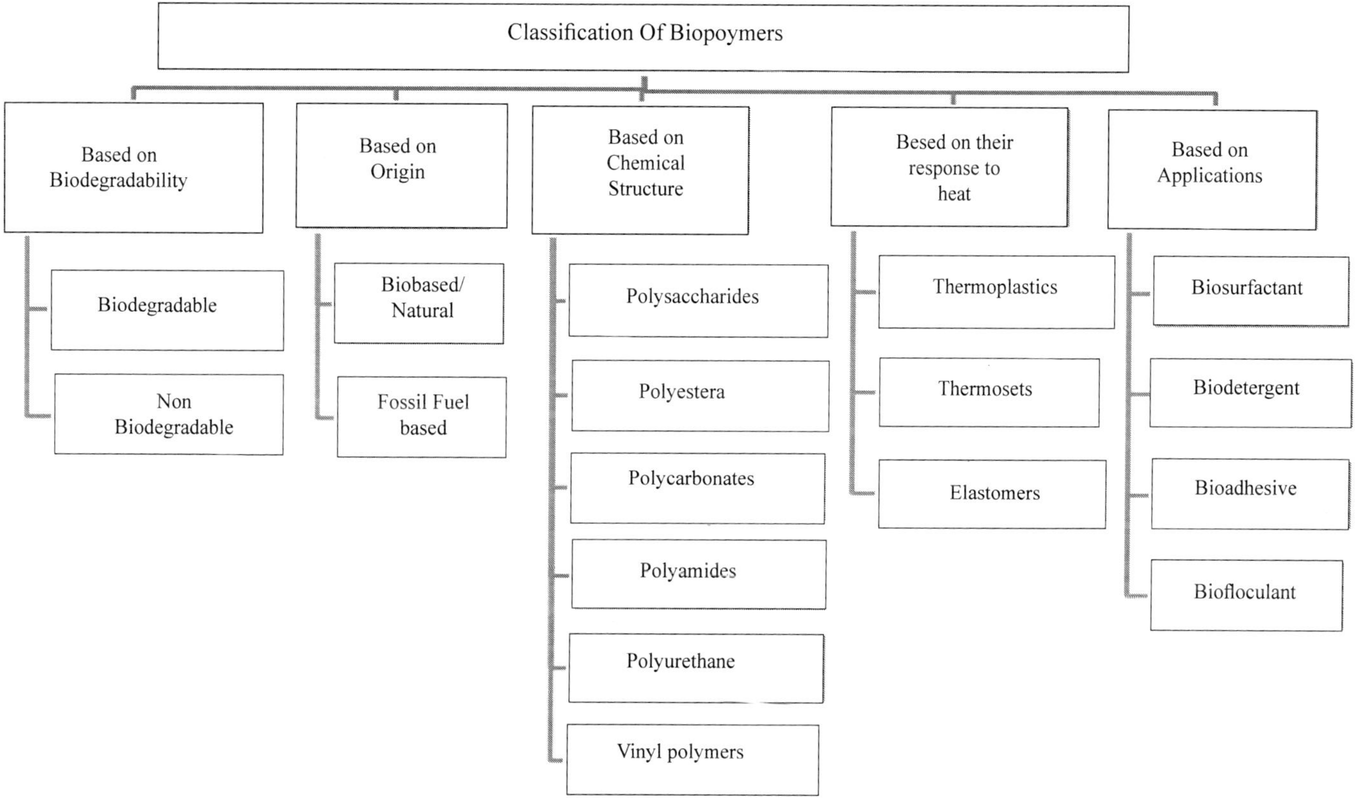

Figure 3: Classification of Biopolymers

Biopolymers can be divided into different categories based on their origin, biodegradability, and chemical structure. They can also be classified according to their response to heat and applications (figure-3).

Based on biodegradability, biopolymers are mainly of two types, Biodegradable and Non-biodegradable[10]. For application in tissue engineering a biomaterial requires properties like biocompatible, non-toxic, providing favorable cellular interactions and tissue development, besides, the properties of mechanical strength, tensile strength etc. For commercial applications like food packaging, biodegradability is important under composting condition so that it does not pose any threat to the environment. Bio based biomaterials are generally extracted from living organisms or are chemically modified from natural renewable resources like, agro-products. Biobased biodegradable biopolymer also includes polymer like PLA (Poly Lactic Acid) which is synthesized from natural lactic acid. On the other hand non-biodegradable biopolymers include various functional biopolymers which are prepared by chemically modifying a bioresource, for example renewable resource's synthetic polyamides from castor oil (polyamide 11), specific polyesters based on bio propanediol, bio polyethylene (bio LDPE, bio-HDPE), bio polypropylene (bio-PP), or bio poly (vinyl chloride) (bio-PVC) based on bioethanol (e.g., from sugarcane), etc. The non-biodegradable biopolymers can be recycled after the end use to reduce the waste generation.

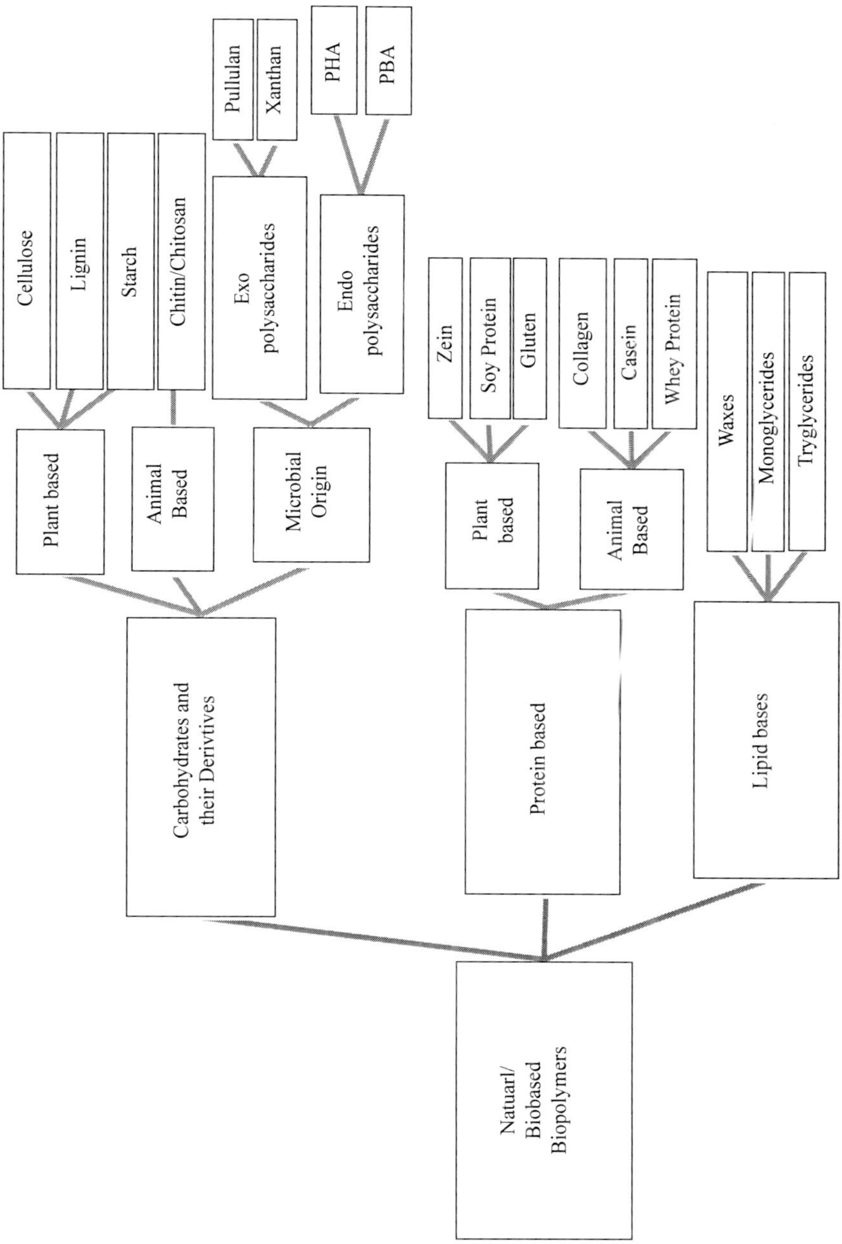

Figure 4: Classification of Natural/ Biobased Biopolymers

According to their origin biopolymers can be divided into two grous: Biobased and Fossil fuel based[10]. As discussed in the earlier section, the biobased polymers have originated from biological system such as plant, animals and microorganisms without altering their structure. They are simply extracted from the natural resources and used after purification. On the other hand, synthetic biopolymers are polymers which are synthesized from a bioresource using other chemical modifiers. Thus the later is not purely a natural compound and the biodegradability of the natural biopolymer may also get adversely affected. The main difference between biobased and fossil fuel based polymers is their sustainability, mainly related to biodegradability. Generally Synthetic biopolymers are used for replacing fossil-fuel based polymers. Biobased biopolymers can be further classified based on their source material, like: a)Sugar based b)protein based and c)lipid based biopolymers (Figure-4). There are a number of fossil fuel based biopolymers available which are biodegradable[1]. The non-degradable bio-based biopolymers are more than biodegradable bio-based biopolymers [3]. The advantages of non-degradable biopolymers or polythioesters are because of their combined effect, they are produced by microorganisms so help to reduce carbon footprints during their production. At the same time they show improved resource efficiency by providing persistence to microbiological degradation. They show durability as well as longetivity and can be act as a substitute of fossil fuel based biopolymers[11]. The market of non degradable biopolymers consist of about 57% of total bioplastics with partially biobased PET at the top followed by biobased polyamides and fully biomass-derived polyethylene[12].

According to the polymer backbone structure biopolymers can be classified into: polyesters, polysaccharides, polycarbonates, polyamides, polyurethane and vinyl polymers[1].

Polysaccharides are complex macromolecules composed when monosaccharide are joint by glycosidic linkage. They are abundant in nature and help in energy reserve and providing structural support in plants. Starch, cellulose chitin and alginates are some commonly known polysaccharides which have biodegradable properties[14] and can be used for packaging applications. These polysaccharides will be covered individually.

Polyamides are polymers composed when two groups are joined together by amide linkage. The monomers like diacids, diamines, aminoacids and lactams can be used for the production of biopolyamides[15]. Polyamides may be either naturally occurring or synthetic in nature. Examples of naturally occurring polyamide include wool, silk etc. Nylon is an example of synthetic polyamide[16].

Polyesters are macromolecule consist of ester as functional group on polymeric main chain. Polyesters are produced from dicarboxylic acids (or any other acid) and diols. Although other forms of polyesters are also present but the term polyester is commonly used for polyethylene terephthalate (PET)[17].

Polycarbonate are thermoplastic molecule composed by a Bisphenol A part and a carbonate group (phosgene is used most commonly). Polycarbonates are popular because of its properties like toughness, high heat capability, optical clarity etc.[18].

Polyurathane are composed from polyisocyanate and polyhydroxy compounds in which urethane group is the major repeating unit. It also contains other groups like ethers, esters, urea and some aromatic compounds. Based on the its properties for different applications polyurethane can be grouped into thermoplastic, rigid, fexible, binders, waterborne, adhesives, sealants, coating and elastomers[19,20].

Polyvinyl alcohol is a vinyl polymer. It is a biodegrable, biocompatible, non-toxic, water-soluble, and synthetic polymer. Polyvinyl alcohol (PVA) can be used commercially for the preparation of blends and composites[21].

On the basis of their response to heat biopolymerts can be categorized into thermoplastics, thermosets and elastomers[10]. Thermoplastics can be softened and melted if heat is applied and processed in the heat-softened state or liquid state[22]. Thermosets are brittle at room temperature and have high cross linkage and low impact resistance and 8787due to this reshaping is not possible after polymerization[23, 24]. Elastomers are loosely cross-linked polymers that are flexible and elastic like rubber[25].

Biopolymers find their applications in different spheres of life, for example biosurfactant, biodetergent, bioadhesive, biofloculant and so on [2]. Biosurfactants (or microbial surfactants) are compounds that can accumulate between fluid phase because of their hydrophilic and hydrophobic moieties. They are used to reduce surface as well as interfacial tension[26]. Bio-based detergents are mixture of surfactants with cleaning properties in dilute solutions[27]. Biodetergents are gaining attention due to their eco friendly nature. Bioadhesion can be understood as the state in which two materials of biological nature are held together by interfacial forces for extended periods of time[28]. Bio adhesive systems are gaining attraction due to its applications in drug delivery which help in reduction of side-effects of non-specific targeting[29]. Bioflocculant can be defined as biodegradable extracellular polymeric material released by bacteria, algae or fungi as the result of cellular proliferation[30]. The applications of bioflocullants include microbial aggregation and biofilm formation. Apart from these biofloculants also acts as emulsifying, stabilizing and viscosifying agents for the of water purification as well as wastewater treatment[31,32].

Characteristics and Applications of Biopolymers

Carbohydrate based Biopolymers

Gum Arabic(GA)

Sources: Gum Arabica, commonly known as Gum Acacia, is an edible bio-based polymer. It is obtained as fiber rich, non viscous fluid from the trees of *Acacia Senegal* and *Acacia seyal*[33].

Chemical Structure: GA is complex polysaccharide. It is a mixture of macromolecules of carbohydrates and proteins of varied size and composition[33]. According to its chemical composition, it mainly consist of carbohydrates (97%) and proteins (<3%)[34]. A slight variation can be observed in composition of Gum Arabica obtained from *A. Senegal* and *A. Seyal* trees. *Acacia Senegal* have high proportion of rahamanose and glucoronic acid[35].

Properties: GA shows high solubility in water. The viscosity of GA is dependent on several factor such as presence of electrolytes, pH of the solution etc. The viscosity of GA solutions is low even at high temperature comparative to other thickners. GA can act as an emulsifier and form. Highly branched structure of GA is responsible for its compactness and small hydrodynamic volume[33].

Advantages: GA has therapeutic effects. For industrial applications GA can act as an stabilizer, thickner, emulsifier and encapsulator[33].

Disadvantages: Prolonged heating above 70 °C can causes precipitation of the proteinous components in Gum Arabic solutions.

Applications: Earlier gum arabica was used in ink and as an adhesive[33]. The industrial applications of GA consist of pharmaceuticals, cosmetics, textiles, encapsulation, food (confectionary, baking, dairy, beverages) etc. [36]. GA has antioxidant and antimicrobial properties which makes it suitable for edible coating[33]. Because of non toxic and biodegradable nature of GA can be considered as better option for packaging industries than other biopolymers[37].

In food packaging, one of the major applications of GA is microencapsulation. In food and pharmaceuticals, it provide extra protection as well as more stable emulsion . GA also have a wide application in fat microencapsulation[38]. Improved structural, gas barrier and thermal properties of GA can be achieved for packaging function by combination of GA and PLA[37].

Cellulose

Sources: Among the major biopolymers, cellulose is majorly found in nature. It occurs mainly in wood, cotton, hemp and some other plant sources. Because of its chemical compostion, cellulose is highly crystalline, insoluble and fibrous in nature[39].

Chemical Structure: Chemically cellulose is a polysaccharide. The monomeric unit of cellulose is unbranched D-glucopyranosyl units (or anhydrofructose unit) joined through β(1-4) linkage. The length of glucan chain depends on the source of cellulose. Through the formation of hydrogen bond, bundles of cellulose forms microfibrils and microfibrils form cellolose fiber[40].

Properties: The characteristics properties of cellulose are associated with its structure. Cellulose fibers naturally show high surface polarity which prevent them to bind with hydrophobic surfaces of petrochemical based polymers[14].

Advantages: The advantages of cellulose include biodegradability, low density, good mechanical properties, flexibility, high durability, and resistance to oil and grease, non toxic, easy chemical modification and transparency[41,42].

Disadvantages: The disadvantages of cellulose include its high water absorbancy and high cost[43] which can limit its application in packaging industry.

Applications: Cellulose are chemically modified for their diverse range of applications for example into carboxymethylcellulose(CMC) and hydroxyethylcellulose(HEC). CMC can be used as a thickening agent, as a binder whereas HEC has major applications in oil industry. Hydropropylcellulose(HPC) is used for the development of flexible films, paper coating etc. A plastic grade material, Cellulose acetate has a potential role in packaging sector[44]. Cellulose has a wide range of applications such as edible films. The films made from cellulose shows great barrier properties to odour , oxygen, moisture and oils[39]. Due to its large surface area to volume ratio, cellulose based biodegradable nanocomposites (CNCs) shows better chemical, mechanical and barrier properties as compared to convectional packaging, thus making the CNCs favorable for packaging applications[43].

Bacterial Cellulose

Sources: Bacterial cellulose or biocellulose is produced from *Acetobacter xylinium.* This microorganism convert sugar into cellulose[45]. Other sources of microorganisms for bacterial cellulose include *Agrobacterium, Azotobacter, Achromobacter, Aerobacter, Alcaligenes, Aerobacter, Dickeya, Pseudomonas, Rhizobium, Rhodobacter, Sarcina, Salmonella*[46,47,48,49]. The sources of carbon can be used for the production of bacterial cellulose are fructose, glucose, mannitol, sucrose, corn steep liquor etc.[50,51,52].

Chemical Structure: It is an exo- polysaccharides and contains linear glucose chains of b-1,4-glucopyranose residues[49]. The molecular formula of bacterial cellulose is $(C_6 H_{10}O_5)_n$ [53].

Properties: It is pure form of cellulose and devoid of hemicelluloses, pectin and lignin[54]. Extrusion of bacterial cellulose from a fixed site leads to its unique properties like ultrafine nanofiber network structure, biodegradability, inert nature, bioadaptibility, hypoallergenity, bioconsistency and chemical stability[55,56,49].

Advantages: Bacterial cellulose show better polymerization and crystallization properties because of absence of impurities like hemicellulose, pectin and lignin associated with plant fibers[57]. Desirable properties of bacterial cellulose include high purity, good binding, coating and thickening properties. The mechanical properties of bacterial cellulose including high water absorption capacity, elastic modulus and high tensile strength are because of its uniform and ultrafine fibrous network structure[58].

Disadvantages: The main limitaition of bacterial cellulose is that price for substrate i.e. sugar is very expensive and it also provide low yield[49].

Applications: Bacterial cellulose is used in paper industries to enhance the strength of paper[59]. It has excellent binding, coating and thickening properties[60]. Because of its structure, it has diverse applications in nano composites, food packaging and active packaging[61].

Chitin and Chitosan

Sources: Chitin, the second most abundant biopolymer next to cellulose, is majorly found in exoskeleton of crustaceans[62].

Chemical Structure: Chemically Chitin is poly (β(1-4)-N-acetyl-D-glucosamine), composed by linkage of glucosamine and N-acetyl glucosamine through β 1→4 glucosidic linkage[63]. Chitosan is a diacetylated derivative of chitin[64].

Properties: The properties of chitosan depend on many factors such as extent of deacetylation, distribution of acetyl groups as well as on the properties of acid used. Chitin is also soluble in mild acids ($pH<6.0$) because of the presence of amino groups[65]. It is insoluble in neutral pH and water[63]. FDA has granted GRAS status to chitosan.

Advantages: Chitosan exhibits high antimicrobial activity. Chitosan based films show good toughness, flexibility and durability. Chitosan based films also show barrier properties against gas and aroma at high concentration[63].

Disadvantages: Chitosan has poor mechanical properties[14]. Chitosan based films are soluble in aquous medium when small quantity of organic acid is present. Due to presence of hydrogen bond chitosan based films are less efficient against humidity[63].

Applications: The major applications of chitosan are in pharmaceuticals, waste water treatment, food processing, active packaging, cosmetics and agricultural sector. Citosan have moisture barrier and antimicrobial properties which can be used for food packaging applications especially when applied as coating[66]. A study by Chien *et al.*(2007)[67] have shown that when mango slice were treated with chitosan solution of different concentrations, better antimicrobial property as well as extended shelf life has been observed. Studies have also suggested that incorporation of chitosan with other polymers can increase the oxygen barrier properties as well as sealability of resulting films and have a wide application in packaging of low moisture foods [3]. Development of chitosan based nanomaterial has also a wide application for future prospective[68].

Starch

Sources: Starch, a plant based biopolymer, constitutes the major portion of various cereals such as rice, corn, wheat etc. and tubers such as manioc, potato etc.[14,69]. It is absent in animal sources.

Chemical Structure: Main components of starch are of amylase and amylopectin. Amylose is composed of α-(1→4)-D-glucopyranosyl monomers, and amylopectin contains the amylose backbone with side units of D-glucopyranosyl linked by α-(1→6) glycosidic bonds[70,14]. The content of amylase and amylopectin depends on the plant source.

Properties: When heated with water starch begin to swell. The film forming properties are mainly because of amylase content of starch[70]. Composites of starch and aliphatic polyesters improves their biodegradability and processability[14],

Advantages: Starch is transparent, tasteless, odorless, and colorless polymer[71]. Advantages of starch include low solubility and barrier properties to oxygen[72]. Both flexible and rigid packaging material can be prepared by starch based biopolymers.

Disadvantages: The polymers of starch and its derivates are brittle and easy to break[71]. Because of hydrophilicity of starch material with poor barrier against water are produced[73]. Without modifications in mechanical properties, starch films can not used for coating purpose[74].

Applications: Due to its low cost and availability, starch can be combined with many other polymers resulting in a diverse range of derivatives. The properties of these derivatives will depend on its building materials, starch and other polymer. Starch have a wide range of application in food packaging health, agriculture, pharmaceuticals and NOVON manufacture[44].

The application of starch in packaging industry is majorly based on edible film and food packaging. It is capable of preserving the characteristics of foods[54]. With certain modifications, a thermoplastic starch can also be produced for making films[73]. Starch is often blended with other polysaccharides and plasticizers to improve its quality and shelf life. This will lead to improved mechanical properties for edible packaging[75].

Lignin

Sources: Lignin is a plant fiber which provides structural support to plants. It is mainly found in tissue of woody and herbaceous plants[44]. It is also a waste product of paper making industry. The content of lignin varies depending upon the sources. Softwoods contain 24–33%, hardwoods 19–28% and cereal straws, bagasse or bamboo about 15–25% lignin[76].

Chemical structure: Lignin has complex amorphous 3-D structure. Primary monomers of lignin include guaiacyl-(G), syringyl-(S) and p-hydroxyphenyl-(H). In the lignin polymer, the primary monomers are joined through C–O bonds of α- and β-arylalkyl ethers[76].

Properties: Lignin provide strength, rigidity and resistance to degradation. It has poor compatability and complex structure[77,78].

Advantages: Lignin is cost effective and renewable with high carbon content. The processing of lignin is energy efficient[77].

Disadvantages: Due of its complex structure and composition as well as poor compatibility, lignin has limited functions in packaging. But at the same time lignin can interact with many other polymers which can improve its mechanical properties, barrier properties as well as resistance to different quality deteriorating factors[78].

Applications: Incorporation of lignin with other polymers improves its functionality for different packaging applications. Many studies have been shown that reinforcement of lignin with other polymers will improve their strength, water uptake and reduce its price. For example; bioplastics made from reinforcement of lignin with cellulose, starch, protein and PLA. As bioplastics, ligin can be used as plasticizer, stabilizer and biocompatabalizer[79]. A Lignin-gelatin based film, prepared with different ratios of lignin and gelatin, has shown improved flexibility and stability[80]. Lignin blended chitosan films provides better antibacterial properties as well as decreased water content and water solubility of resulting films[81].

Poly Lactic Acid(PLA)

Sources: PLA, a biodegradable polymer of lactic acid, is produced through the process of fermentation. Lactic acid bacteria convert stach into sugar then ferment it into lactic acid[82].

Chemical Structure: PLA is an aliphatic polyester of lactic acid based monomers[83]. The synthesis of monomers of PLA includes polycondensation by ring-opening polymerization of lactide[84]. The properties of PLA are strongly related to the content of lactic acid isomerization[49].

Properties: The properties of starch are comparable to LDPE with exception of rigidity which is higher. Hydrolysis of the ester linkage leads to hydrosis of polymer[14]. As compared to other biopolymers, PLA shows very good properties such as low toxicity, biodegradability, sustainability, biocompatibility, high mechanical strength and high process ability[85].

Advantages: PLA based films are cost effective, compostable and burnable and show good printability, heat sealability and resistance to water and oil.

Disadvantages: Stiffness, low permeability and proneness to hydrolysis are several limitation of PLA.

Applications: To improve its quality and make it cost effective, PLA is sometimes blended with other polymers[86]. PLA-PGA copolymer, with glycolic acid, has been approved by FDA and mainly used in medical and agricultural sector[44]. The antimicrobial activity of PLA enables its application for contact packaging in foods. The main applications of PLA are in agricultural films, single use cutlery, biomedical devices packaging and automotive industries.

Polyhydroxyalkanoate (PHA)

Sources: Polyhydroxyalkanoates (PHA), a biopolymer from the family bacterial polyesters, are produced by fermentation of lipids and sugar by bacterial species**[87].** There are various bacterial species which are used for the production of PHA .Some important species include *Azotobacterium sp., Alcaligenes latus, Alcaligenes eutrophus, Bacillus sp., Rhizobium sp., Methylobacterium sp.* and *Pseudomonas oleovorans*[49].

Chemical structure: Based on number of carbon atoms, two variety of PHA are there, first is scl-PHA, made of monomers with 4-5 carbon atoms and mcl-PHA, made of monomers with 6-14 carbon atoms[88,49].

Properties: The thermal and mechanical properties of PHA are similar to PP, a synthetic polymer[89]. The desirable properties of PHA consist of good barrier properties to flavor and odour, printability, heat sealability, barrier to oil and grease, good tensile strength etc.

Advantages: PHA polyesters have several advantages such as biodegradability, biocompatibility and it is cost effective and made from renewable sources[90]. Because of their biodegradable nature it helps in displacement of petroleum and minimization of green house gases[91].

Disadvantages: Dispite of having several advantages, PHA have several limitations such as, the films produced by PHA are stiff and brittle. Low viscosity and thermal instability of PHA films also limit its applications[49].

Applications: Blended films of PHA with glycerol, PVC, PVA show improved mechanical properties[92]. The applications of PHA in food and packaging industries include as a biodegradable packaging material, coating and laminates, fiber, container, bottles, sheet, films etc[87]. The applications of PHA also include production of disposable feminine hygiene products, disposable food containers, shampoo bottle etc.[49].

Poly-beta-hydroxybutyrate (PHB)

Sources: Polyhydroxybutyrate is the first discovered polymer of PHA family. It is most widely produced by bacteria *Alcaligenes eutrophus*[90].

Chemical Structure: PHB belongs to the polyhydroxyalkanoate family of biopolymers. It is polyester of 3-hydroxybutyric acid or 4-hydroxybutyric acid[25].

Properties: It is a partially crystalline material with high melting point. The properties of PHB are very similar or comparable to synthetic petrochemical based plastics. It is totally biodegradable, brittle, not soluble in water and has good barrier properties[93].

Advantages: It is extruded, molded and spun into fiber to made into films. PHB can be processed into varied structure i.e fibers, sheets, films, spheres etc. PHB show several advantages like gas barrier properties, low toxicity and degradability[1]. To reduce its brittleness, PHB can be mixed or blended with other biopolymers such as PCL, PVOH, PLA and starch. The incorporation of nanofillers may also increase the toughness of PHB[90]. PHB degrades into D-3 Hydroxybutyrate, which is naturally present in human blood.

Disadvantages: Because of high crystallinity, PHB shows poor impact strength, stiffness and brittleness. These characters of PHB leads to very poor mechanical properties and low extension at breakage which limits its potential applications[94]. Degradation of PHB is also slow as compared to other biopolymers.

Applications: The limitations of PHB can be prevented by formation of polymer or plasticizer based blends of PHB[94]. Mainly PHB is used in biomedical applications in tissue engineering and drug delivery.. PHB is also used as a packaging material and in disposable products[90].

Pullulan

Sources: Pullulan, a natural biopolymer, is obtained from *Aureobasidium pullulan***[49].** The different carbon sources used in the production of pullulan include glucose, lactose, maltose, sucrose, jaggery, and peat hydrolysate[95].

Chemical Structure: Pullulan is an extracellular, water-soluble polysaccharide. It has unbranched, linear structure with molecular formula $(C_6H_{10}O_5)_n$. It contains repeating units of maltotriose linked together with α-1,6 glycosidic bonds. The glycosidic bonds of pullulan are responsible for its characteristics properties[49].

Properties: Pululan is transparent, insoluble in organic solvent, non toxic, free from taste and odour[44]. Desirable properties of pullulan include low viscosity, readily soluble in water and excellent adhesive characteristics, which makes it more suitable for cosmetic products like hair conditioner, shampoo etc. It is also used as an active ingredient in antiaging products[54].

Advantages: Pullulan show transparency, gloss, hardness, strength and toughness resembling to polymers like PVC and PS (Leathers, 2003). The elasticity of pullulan is better than convectional plastics[95]. It can be easily moulded into to convert them into pouch bags. The antimicrobial and barrier properties of these films against moisture and oxygen help in shelf life extension and rancidity reduction in vegetable oils. It forms tasteless and colorless films that are resistant to temperature and grease with good printability and heat sealability[96].

Disadvantages: The limitations of pullulan include high water solubility, The disadvantages Include poor water resistance, brittleness, difficult to process by conventional melt processing techniques[97].

Applications: Pullulan can be modified using chemicals to make it less soluble or insoluble in water. By compression molding, pullulan can be converted into plastic resembling articles so it has wide potential as industrial plastic. It can be used as food additive for bulking and texturizing effects.. In food industry, pullulan can be used for making edible water soluble film for food products. Pullulan compounds are also used as drug and gene carrier for biomedical applications[44]. Pullulan can also be used as stabilizer, binder and low viscosity filler in beverage and bakery industry[49].

Alginates

Sources: Alginates, extracted from brown seaweed, is a structural constituent of brown algae and carrageenan[14].

Chemical structure: Alginates are salt of alginic acid which is a linear copolymer of L-guluronic acid and D-mannuronic monomers[14].

Properties: It have excellent colloidal and gelling properties as well as b NJ,Marrier properties for oil and grease. Alginates are soluble in water, in both hot and cold but insoluble in solvent[54]. Alginates can react with polyvalent metal cation[98].

Advantages: Films produced from alginate provide barrier to greases and oils and oxygen. Alginate coatings also act as a sacrificing agent to preserve meat and prevent oxidation because of its oxygen barrier properties[14].

Disadvantages: Ion leaching of alginate films leads to formation of unstable film. Other limitations of alginate include its limited control of mechanical properties amd poor water resistance[98,99].

Applications: In food industry alginates are used for production of icecream and other dairy products and for the formation of edible coating for strawberry, kiwi etc[49]. The applications also include restructured foods such as cocktail berries, onion rings and meat products[100]. It can be also used in sizing/ coating paper to enhance smoothnes and printability, textile industry, and in biomedical applications for encapsulation and drug delivery[44].

Protein Based Biopolymers

Proteins are polymer made up of amino acids. Various proteins are present naturally in living organisms as the structural material and some of the structural proteins are chemically synthesized to form a polypeptide. Being natural or synthetic, these proteins are physically or chemically modified for novel applications in biochemistry and food industry. Apart from its highly nutritional and sensory characteristics, protein based edible films have property to protect the food from its environment. Most of the research related to protein is limited to high-tech applications like bioceramics, elastomer, electro-optical materials and adhesive due to high production cost[44].

Proteins of Animal origin

Collagen

Sources: Collagen is the most abundant protein in human and found in skin, muscle, bones and tendon to provide strength and structure[2]. It is the primary protein of animal connective tissue[82].

Chemical structure: Collagen is made up of three polypeptide subunits with similar amino acid composition. Each polypeptide consists of 33% glycine, 25% hydroxyproline and 25%, proline and lysine in abundance. Each polypeptide contains approximately 1050 amino acids[101,102].

Properties: Collagen properties are similar to plant cellulose. Collagen degrades into its corresponding amino acids in the presence of enzymes collagenases and metalloproteinases within the body. The enzymatic treatment or cross linking can significantely change the rate of degradation of collagen[102].

Advantages: Collagen is available in abundance and non-antigenic in nature.The desirable properties of collagen include its biocompatibility, bioresorbability, biodegradability and non-toxicity. Collagen has minimal expressibility and high tensile strength. It is easily modifiable and can be formulated in various forms. It is also compatible with synthetic polymers[103].

Disadvantages: The allergic effects of collagen to several people are also responsible for limiting its applications[104]. One of the major drawback of collagen is that during manufacturing higher quantity of solvent is consumed and price per gram of solid collagen is also high. Heating of collagen leads to its denaturation[105]. The mechanical strength of collagen is also poor which limits its applications for packaging[106,103].

Applications: Applications of collagen are mainly related to biomedical. It can also be processed into powder, foams, tube, sheets, dispersions, sponges nanofibrous matrices, etc.[54, 102]. Collagen film or casings can be made to preserve meat and meat products from oxidative deterioration, shrinkage loss and to increase juiciness[14].

Gelatin

Sources: Gelatin is obtained from partial hydrolysis of collagen[82].

Chemical structure: Gelatin is a mixture of peptides and proteins[107]. Gelatin contains large number of proline, glycine and 4-hydroxy proline residues[108]. Depending on the type of hydrolysis, generally two types of gelatin are obtained- type A & type B. Acid hydrolysis will results in type-A and alkaline hydrolysis will results in production of type-B gelatin[109].

Properties: Gelatine is a partially cleaved, denatured collagen with broad distributions of molecular weight[105]. It is translucent and permeable to water[14].

Advantages: Gelatin can prevent lipid oxidation and dehydration. It is insoluble in non polar solvents and alcohol. Gelatin also exhibits barrier properties against light and oxygen[102].

Disadvantages: The major drawback of gelatin is that it is highly permeable to water vapour[105]. Poor mechanical properties of gelatin are responsible for its limited application in packaging industry[108].

Applications: Gelatin is used in food industry as emulsifying, stabilizing and foaming agent. It can also be used for the formation of biodegradable films[82]. In drug delivery, it helps in slower or no release of drug. Modification of gelatin can be done with nanoparticles and biomolecules for wide applications. In biomedical science gelatin can be used for transfer of genes and wound dressing [2]. Gelatin coating are used for meat and meat products from very old time to prevent microbial spoilage and dehydration[14].

Whey protein

Sources: Whey protein is a by-product obtained from cheese industry. It consist of whey protein isolates (which is purer form of such proteins) and whey protein concentrates[110,82]. It is about 20 percent of total milk protein[102].

Chemical structure: Whey protein has globular strucure. There are various proteins with diverse functionality which constitute whey protein. The primary proteins include α-lactalbumins, β-lactoglobulins, immunoglobulins, bovine serum albumin and protease peptones. α -lactalbumins, β -lactoglobulins and bovine serum albumin content in whey protein is about 57, 19 and 7 percent respectively[112,102].

Properties: Whey protein based films are biodegradable and capable to control, carbon dioxide, lipid, flavour and aroma transfer[113,114,102].Whey proteins are good barriers to oxygen and have a moderate permeability to moisture[115]. Whey proteins show solubility at pH 4.6[112].

Advantages: Whey protein based films are transparent and show barriers properties against oxygen which makes it suitable for formation of elastic films[116]. It can also be used as a raw material for biodegradable packaging.

Disadvantages: Whey protein based films are brittle & stiff and show poor moisture barrier properties.

Applications: Films made from whay proteins acts as a protective against uptake of oxygen and rancidity. Incorporation of ascorbic acid into these films will further improve oxygen scavenging action and reduce the oxygen transmission rate to protect from oxidative rancidity[117,14]. Whey protein is also used for the manufacturing of edible films[102].

Casein

Sources: Casein is the primary protein of milk and constitute about 80 percent of total milk protein[118].

Chemical structure: Casein is a phosphoprotein which differ in their structure (primary, secondary and tertiary) and molecular weight. It can be separated into different electrophoretic fractions such as α_1-casein, α_2-casein, β-casein and Ϗ-casein[102].

Properties: Casein is thermal stable, biodegradable and non toxic in nature and it has ability to bind small molecules and ions which make it a good material for biodegradable packaging film and as an adhesive[119]. Casein has ability to form film without further processing in aqueous medium due to strong interchange cohesion[118].

Advantages Casein can be used for the manufacturing of edible films[14] Casein based films are water insoluble but they can absorb a large quantity of water. They are opaque with good mechanical properties[118].

Disadvantages: Most casein based films are sensitive to moisture which affects its mechanical and barrier properties[120].

Applications: For the modification of casein based films, physical and chemical treatment can be done to improve its functionality. Blending of casein with other biopolymers can also be done for the same[120].

Proteins of Plant origin

Wheat protein

Sources: Gluten is a group of protein majorly found in cereals. Many times gluten is correlated with wheat proteins. Gluten represents about 70-80% of wheat storage proteins[102].

Chemical structure: Gluten consists of two storage proteins, gliadin and glutenin[118]. Gliadin is the soluble fraction of prolamins and glutenins are the insoluble fraction of glutelin. These fractionsare responsible for constituting gluten biopolymer[121].

Properties: Gluten, wheat protein is known for its viscoelastic properties which make it suitable for film making. Other properties of gluten include opacity, water solubility, mechanical properties and barrier properties[122]. Gluten based films are shiny and resistant to water. They are water insoluble but able to absorb water[123].

Advantages: Gluten proteins are capable to form continuous viscolastic network. Gluten is biodegradable, cost effective and available in abundance**[124].** Rapid biodegradability of gluten can be an advantage for packaging of some non-food products. The mechanical properties of gluten are also comparable with plastic based rubber[125].

Disadvantages: Washing can produce high grade gluten but this is a disadvantage because of consumption of high amount of water[124].Some people shows sensitivity to gluten (gluten allergy and intolerance) which can limit its applications.

Applications: Gluten can be processed to form edible film[14]. Gluten can be blended with other material to improve its properties for film making[61]. As compared to plastic films, gluten based films show high water-vapour permeability and low permeability to oxygen and carbon-dioxide[61]. Hydrolyzed wheat protein, obtained from hydrolysis of gluten, can be used as a stabilizing and emulsifying agent in food and also added in cosmetics as a active ingredient[54]. Other fields of applications of gluten include biodegradable plastic films, surface coatings on paper, bakery industry, detergents, cosmetics, etc.[126].

Zein

Sources: Zein is the naturally occurring storage protein of corn and obtained from industrial processing of corn. It represents more than 50% proportion of total endosperm protein[102].

Chemical structure: Zein is made up from prolamin fraction of corn proteins[14]. The three primary protein fraction of zein include α-zein, β-zein, and γ-zein which can be differentiated based on differences in their solubility[127, 102].

Properties: Generally zein do not show solubility in water. It is soluble in alcohol but not soluble in anhydrous alcohols[14]. It has moisture, lipid and oxygen barrier properties[117].

Advantages: The advantages of zein protein include its biocompatability and biodegradability. Zein based film has water barrier properties. It also shows barrier properties against oil and grease. Plasticization of zein can improve its property and can significantly change its rheological properties[128].

Disadvantages: Generally zein is not water soluble but but it become soluble at very high or very low pH[14]. It shows poor mechanical properties and water vapour barrier[129]. Zein based film are also tough and brittle.

Applications: Zein can be used for the development of various thermoplastic products because of its barrier properties to oxygen and hydrophobic nature[82]. According to Hefft[61] in his research published in Journal of Applied Packaging research, films made from zein protein have comparable barrier properties to partially etherified cellulose. Zein can also be used for formation of edible films[14]. Zein can be used as finishing agent for providing surface gloss for nuts, confectionery and other foods products[117,14]. Zein based films can be used to protect food having short shelf life[123].

4.2.7. Soy protein

Sources: Soy protein is obtained from soyabean. It is commercially available as soy flour, soy protein isolates and soy protein concentrate[130].

Chemical structure: About 90% of soy proteins are globulins and can be fractioned on the basis of their sedimentation coefficients into 2S, 7S, 11S and 15S[121]. Soy proteins are tightly folded proteins by their conformation[131].

Properties: The emulsifying and texturizing properties of soy protein make it suitable for food sector[132]. Other important functional properties of soy protein include its biodegradability, solubility, cohesiveness, adhesiveness, foaming, dough and fibre formation etc.[102].

Advantages: Soy protein based films show improved transparency, smoothness, clarity and flexibility comparative to other plant based films[133,102]

Disadvantages: Soy protein based films have poor mechanical properties and highly sensitivity to moisture. Some people are allergic to soy proteins which is a major disadvantage for its application in food and packaging industry[14].

Applications: The application of soy protein includes manufacturing of plastic material. It also acts as an adhesive[134]. It is also used for edible coating and as an active ingredient in cosmetic products[14,54]. Applications of soy polymers also include antimicrobial films, paper coating, dyes and glues[118].

Lipid Based Biopolymers

Sources: Natural lipids are obtained from various plants and animal sources. Plant sources of lipids include fat from cocoa, nuts, oilseeds as well as some cereals and pulses. Animal sources of lipids include waxes, butterfat, lard, tallow, cholesterol etc. Lipids can also be synthesized. Some example of synthetic lipids include paraffin and pertolium wax.

Chemical Structure: Lipids are nonpolar molecule. The structure of lipids varies widely but generally consists of hydrocarbon chains. Specific types of lipids can also include other atoms like oxygen and phosphorus[136]. When interact with water, lipids forms micelles, monolayers, bilayers, membranes or vesicles because of hydrophobic effect[136].

Different types of Lipid based biopolymers: Lipid based biopolymers can be grouped into two classes which are polar and non polar[137]. The non polar lipids contains waxes while the polar lipids consist of monoglycerides, triglycerides and some others lipid based molecules like phospholipids, cholesterol, fat soluble vitamins which are not very Important for packaging function .

Waxes

Waxes include natural waxes such as beeswax, candelilla wax, carnauba wax, rice bran wax, and synthetic waxes such as petroleum wax and paraffin[14]. Waxes are example of non polar lipids. Waxes are not soluble in water because of absence of hydrocarbons in their structure. They have poor spreadability so do not spread on the surface for the formation of monolayer[138]. Because of these properties waxes act as effective barrier against water vapour[137].

Triglycerides

Triglycerides are example of polar lipids. There is slight difference in the functionality of long and short chain triglycerides. Short chain triglycerides show partial solubility in water while long chain triglycerides are insoluble in water but it has the ability to spread at the interface of water for the formation of stable monolayer[136]. Addition of stearyl alcohol and palmitic, lauric, stearic acids to edible films will improve its moisture barrier properties for packaging function[137,138].

Monoglycerides

The solubility or insolubility of monoglycerides in water is a function of their chain length. Some monoglycerides are insoluble in water but they possess the ability to solublize water in their hydrophilic part. These monoglycerides are sparingly soluble in non polar solvents (like hexane) in the absence of water. Because of these properties they act as excellent emulsifiers. When the water content is less they form inverted micelles with polar groups towards inside. And when the water is in excess, normal micelles are formed with polar heads toward aqueous phase[136]. Monoglycerides are used for stabilizing emulsion and to enhance adhesion in the bilayer film[137].

Acetylated monoglycerides

In Acetylated monoglycerides, acetic acid is bound with monoglyceride for emulsification properties[139]. Due to their plasticizing properties acetylated monoglycerides are often used in edible packaging for coating of frozen foods[137].

Properties: Lipids are better moisture barrier as compared to proteins and polysaccharides. Saturated forms of fatty acids have higher melting points and lower rate of moisture transfer as compared to the unsaturated forms. Partial water solubility can be seen in case of long chain lipid molecules. Barrier properties of lipids are dependent on various factors like length of fatty acid, its chemical structure, physical properties, partical size, interaction with other components (protein or polysaccharides) of film etc[140].

Advantages: Lipid based films show good stability and hydrophobic properties so they are effective barrier against moisture[136]. Lipids can provide gloss to packaging material[14]. Lipid based biopolymers can be used as an effective edible coating with incorporation of antimicrobial substances which helps to enhance the shelf life of the food product.

Disadvantages: The applications of lipid based biopolymers are limited because of their inability for the formation of film due to weak mechanical properties[118]. Incorporation of lipids provide good barrier properties but sometimes lipid based films are too brittle, opaque and inflexible[137]. Lipids donot form continuous and cohesive matrices[140]. Lipid based films leads to waxy sensation so having an undesirable effect on sensory attributes[141]. Application of lipid based coating are difficult due to their thickness and greasy surface[14].

Applications: Lipids can be incorporated with proteins or polysaccharides for the formation of biocomposites with improved barrier properties[142]. Some studies have been shown that incorporation of sunflower oil to chitosan films lead to reduction of metamyoglobin content of meat[143] while addition of beewax in chitosan based films leads to decrease in rate of respiration of strawberries[144]. Another study has been shown that coating of sausages with lipids will provide moisture barrier properties and enhance the stability of the film[145]. The lipid based coating on dairy products reduces the microbial contamination and water loss during its storage[111]. In packaging of bakery products lipid based films prevent the transfer of water and helps to enhance the shelf life of the product[146].

General Applications of Biopolymers

Biopolymers have a wide range of applications(Figure-5) such as commodity clothing, packaging, industrial plastics, absorbent materials, medical textiles, biosensors, biofilters and biobarrier etc[147,148]. Blending of biopolymers is also done for improvement of its properties. We can blend a biopolymer with another biopolymer- natural or synthetic, which may be biodegradable or non biodegradable. Biocomposites are often made by reinforcement of biopolymer with another material such as mineral or natural fibre. This will lead a new material with better properties[149].

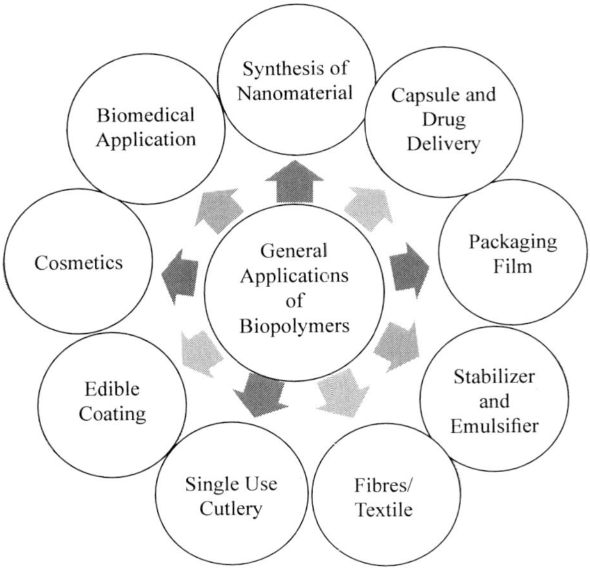

Figure 5. General Applications of biopolymers

Because of their biocompatibility and biodergradability, biopolymers can be utilized for various applications. Some applications of biopolymers are given below:

As a packaging material

Bio-based polymers are generally used to replace the petrochemical based packaging material which further leads to reduction of overall carbon footprint related to these packaging materials[150].

Rigid bioplastics such as PLA, bio-based PE are widely used in the packaging of beverage bottles, for the packaging of cosmetics such as lipstick and cream etc. Flexible packaging tray and films are also used for improving shelf life in fresh fruits and vegetables[9]. Now active packaging is also done with the help of biopolymers. Chitosan show some antimicrobial properties which is incorporated in packaging material for preservation against various microbes**[2].** Some biopolymers like starch, when mixed with plasticizers are used in the formation of thin films for packaging function[151].

Food industry

Biopolymers have found a wide interest in single use cutlery, food packaging material, encapsulation of functional foods and edible coating of foods etc. Biopolymers are also used as a stabilizer, adhesive/ binding agent, emulsifier and thickening agent in food industry[2].

Agriculture and horticulture

Biopolymers are generally used for minizing the use of pesticides for production of pure foods or organic farming. Another potential applications of biopolymers in the field of agriculture and horticulture are films for banana bushes for protection against dust and environment, plant pots for cultivation or propagation, fertilizer rods etc. Because of biopolymers, flower bulbs which can be planted with their packaging into the soil are also available commercially. This packaging disperses after the planting [9].

Synthesis of nanomaterial

With the help of nanotechnology, biopolymers are majorly used as a capping agent or reducing agent in synthesizing various nanoparticles. Various biopolymers such as PVA, chitosan, cellulose are used for this purpose[2].

Several nanomaterials(for eg, silver nanoparticle) also show some antifungal and antibacterial properties which can be utilized to provide microbial safety in drinking water. A new polymer, *chitosan* is also used in water treatment as a flocculant[2].

Biomedical applications

These days biomaterials have found wide interest in medical devices, pharmaceutical carriers, drug delivery system and tissue engineering[152,153,154]. For eg.*gelatin* is extensively used as an adhesive and wound healing[2].

Other applications

Other applications of biopolymers include fiber/ textile use eg, technical textiles such as fleeces; consumer goods such as organic waste sacs, cotton buds and disposable gloves, in building and construction, automotive and electronics etc.[155].

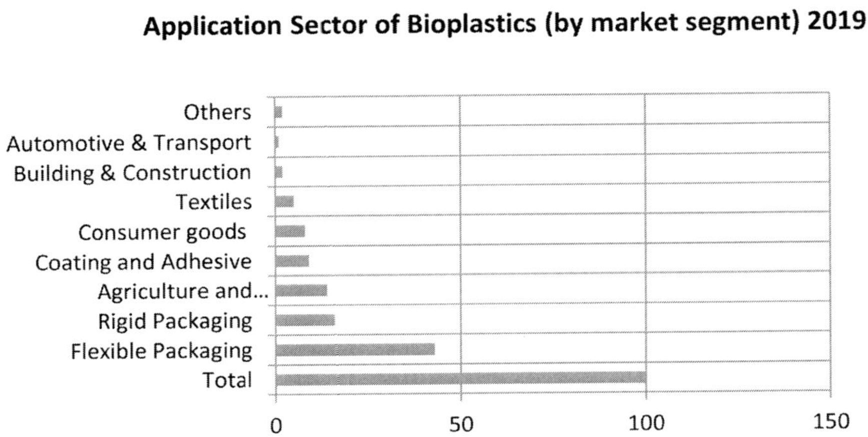

Figure-6. Market segment of bioplastics [9]

Future Trends and Conclusion

Biopolymers encompass a wide range of applications, finding use in various fields from nanotechnology-enabled controlled release to edible drug delivery systems. They have become a new building block in macromolecular chemistry. The food industry is embracing a new trend of using renewable and recyclable agricultural products for coatings, edible films, and packaging films. Similarly, the packaging industry prioritizes reducing raw material usage and cost, while minimizing waste generation.

However, despite their versatility, biopolymers currently represent a small portion of the commercial market. Several challenges hinder wider adoption, such as the high cost of biopolymers compared to traditional materials, limited raw material availability, and the complexities of segregation and processing. Additionally, developing nations often lack clear regulations for biopolymers, and comprehensive life cycle assessments are still needed. Many biopolymer studies have focused on laboratory-scale applications. Future research should prioritize utilizing biopolymers for material and energy generation to save fossil fuels and reduce carbon footprints. Life cycle analysis, processability improvements, and better finishing techniques are also crucial for biopolymers to become commercially viable. A sustainable and interdisciplinary approach is essential to successfully implement and commercialize biopolymer-based products.

References

1. Aadil, K.R., Barapatre, A. & Jha, H. (2016), Synthesis and characterization of Acacia lignin-gelatin film for its possible application in food packaging, Bioresources and Bioprocessing, 3:27. DOI: 10.1186/s40643-016-0103-y

2.Ab. Halim, N.I. (2010), Synthesis of bacterial cellulose by Acetobacter Xylinum sp. using pineapple pith for biocomposite application. http://umpir.ump.edu.my/id/eprint/3291/1/CD5878_NURUL_IFFAH_BTE._AB._HALIM.pdf

3.Abu Tawila Z., Ismail S., Dadrasnia A., Usman M. (2018), Production and characterization of a bioflocculant produced by Bacillus salmalaya 139SI-7 and its applications in wastewater treatment. Molecules. 2018;23:2689. DOI: 10.3390/molecules23102689

4.Akindoyo, John & Beg, Mohammad & Ghazali, Suriati & Islam, Muhammad & Jeyaratnam, Nitthiyah & Ar, Yuvaraj. (2016). Polyurethane types, synthesis and applications-a review. RSC Adv.. 6. 114453-114482. DOI: 10.1039/C6RA14525F.

5.Amin MCIM, Halib N & Ahmad I, 2010, Unique stimuli responsive characteristics of electron beam synthesized bacterial cellulose/acrylic acid composite, Journal of Applied Polymer Science, 116, 2920–2929

6.Andreeßen, C., Steinbüchel, A (2019). Recent developments in non-biodegradable biopolymers: Precursors, production processes, and future perspectives. Applied Microbiology and Biotechnology 103, 143–157. DOI: https://doi.org/10.1007/s00253-018-9483-6

7.Ansari, S. and Fatma, T. (2014), Polyhydroxybutyrate - a Biodegradable Plastic and its Various Formulations, International Journal of Innovative Research in Science, Engineering and Technology, Vol. 3, Issue 2, February 2014, ISSN: 2319-8753, Page No- 9494-9499

8.Argos, P., Pederson, K., Marks, M.D., Larkins, B.A. (1982), Structure model for maize zein proteins, J Biol Chem 257:9984–9990

9.Arvanitoyannis, I.S. (2013), Waste Management for Polymers in Food Packaging Industries, From book Plastic Films in Food Packaging, Chapter 14, page 249-310, doi: https://doi.org/10.1016/B978-1-4557-3112-1.00014-4

10.Ashraf, M.A., Batool, S., Ahmad, M., Sarfraz, M. & Noor, W.S.A.W.M. (2016), Biopolymers as biofilters and biobarriers, From book Biopolymers and Biotech Admixtures for Eco-Efficient Construction Materials, Chapter 17, Pages 387-420. https://doi.org/10.1016/B978-0-08-100214-8.00017-8

11.Ata, R., Aladdin, A., Othman, N., Malek, R.A., Leng, O., Aziz, R. & El Enshasy, H. (2015), Recent applications of polylactic acid in pharmaceutical and medical industries, Journal of Chemical and Pharmaceutical Research, ISSN 0975-7384, 2015(712), page 51-63. https://www.researchgate.net/publication/292990733

12.Augustine, R., Rajendran, R., Cvelbar, U., Mozetic, M. & George, A. (2013), Biopolymers for Health, Food, and Cosmetic Applications, From book Handbook of Biopolymer-Based Materials: From Blends and Composites to Gels and Complex Networks, Chapter 27, page 801-851 . DOI: 10.1002/9783527652457.ch27.

13.Aydin, F., Kahve, H. & Ardic, M. (2017). Lipid-based edible films. The Journal of Scientific and Engineering Research. 4(9): 86-92. https://www.researchgate.net/publication/320373382_Lipid-based_edible_films

14.Bertuzzi, M.A., Armada, M. and Gottifredi, J.C. (2007), Physico-chemical characterization of starch based films, Journal of Food Engineering, volume. 82, page. 17–25, DOI: 10.1016/j.jfoodeng.2006.12.016

15.Bielecki S, Krystynowicz A, Turkiewicz M & Kalinowska H, 2005, Bacterial cellulose. In- Biotechnology of Polymer: From Synthesis to Patents, A Steinbuchel (ed), Munster, Germany: Wiley-VCH, Verlag GmbH, pp 381–434

16.Bioplastics News, Alex Barret (July,5, 2018), The History and Most Important Innovations of Bioplastics, Retrieved from URL https://bioplasticsnews.com/2018/07/05/history-of-bioplastics/

17.Bourtoom, T. (2008) Edible films and coating: characteristics and properties, Int Food Res J 15(3): 237–248

18. Brahmbhatt, Devarshi. (2017). Bioadhesive drug delivery systems: Overview and recent advances. International Journal of Chemical and Life Sciences. 6. DOI: 10.21746/ijcls.2017.3.1.

19. Brunelle and Korn (2005), Advances in Polycarbonates, ACS Symposium Series; American Chemical Society: Washington, DC.DOI: https://pubs.acs.org/doi/pdf/10.1021/bk-2005-0898.ch001

20. Callegarin, F., Gallo J-A. Q., Debeaufort. F. And Voilley A. (1997). Lipids and Biopackaging. Journal of the American Oil Chemists' Society. Vol. 74 No: 10 Pages: 1183-1192. https://link.springer.com/content/pdf/10.1007/s11746-997-0044-x.pdf

21. Castro-Rosas, Javier & Cruz-Gálvez, Andrés & Gomez-Aldapa, Carlos & Cortés, Reyna & Guzmán-Ortiz, Fabiola & Rodriguez, Maria. (2016). Biopolymer films and the effects of added lipids, nanoparticles and antimicrobials on their mechanical and barrier properties. International Journal of Food Science & Technology. https://www.researchgate.net/publication/303753105_Biopolymer_films_and_the_effects_of_added_lipids_nanoparticles_and_antimicrobials_on_their_mechanical_and_barrier_properties

22. Cauchie, H.M. (1997), An attempt to estimate crustacean chitin production in the hydrosphere, International conference; 7th, Chitin and chitosan; 1997, page 32-39; Lyon; France. https://www.tib.eu/en/search/id/BLCP%3ACN027459562/An-attempt-to-estimate-crustacean-chitin-production/

23. Cha. D. S. and Chinnan. M. S.(2004). Biopolymer-Based Antimicrobial Packaging: A Review. Critical Reviews in Food Science and Nutrition. Vol. 44 No: 4 Pages: 223-237.

24. Chassenieux, C., Durand, D., Jyotishkumar, P. & Thomas, S. (2013), Biopolymers: State of the Art, New Challenges, and Opportunities, From book Handbook of Biopolymer-Based Materials: From Blends and Composites to Gels and Complex Networks, First Edition, Chapter 1. https://doi.org/10.1002/9783527652457.ch1

25. Chaudhary, D., Dong, Y. & Kar, K. (2010), Hydrophilic plasticized biopolymers: Morphological influence on physical properties, Materials Letters, Volume 64, page 872-875. DOI: 10.1016/j.matlet.2010.01.048.

26. Chawla PR, Bajaj IB, Survase SA & Singhal RS, 2009, Microbial cellulose: fermentative production and applications, Food Technology and Biotechnology, 47, 107–124

27. Chen, C.-T., Chen, K.-I., Chiang, H.-H., Chen, Y.-K. and Cheng, K.-C. (2017), Improvement on Physical Properties of Pullulan Films by Novel Cross-Linking Strategy. Journal of Food Science, 82: 108-117. DOI:10.1111/1750-3841.13577

28. Chen, H., Wang, J., Cheng, Y., Wang, C., Liu, H., Bian, H., Pan, Y., Sun, J. & Han, W. (2019), Application of Protein-Based Films and Coatings for Food Packaging: A Review, Polymers, Volume 11, DOI: 10.3390/polym11122039.

29. Chien, P., Sheu, F., & Yang, F. (2007). Effects of edible chitosan coating on quality and shelf life of sliced mango fruit. Journal of food engineering, 78, 225-229. doi: 10.1016/j.jfoodeng.2005.09.022

30. Conca KR & Yang TCS, 1993, Edible food barrier coatings, In- Biodegradable Polymers and Packaging, C Ching, DL Kaplan & EL Thomas (eds), Lancaster: Technomic, pp 357–369

31. Credou, J. and Berthelot, T., (2014), Cellulose: From biocompatible to bioactive material. J. Mater. Chem. B, 2, 4767–4788

32. David, Anu & Day, James & Shikanov, Ariella. (2016). Immunoisolation to prevent tissue graft rejection: Current knowledge and future use. Experimental Biology and Medicine. 241. 10.1177/1535370216647129.

33. Deman JM (1999), Principles of Food Chemistry, 3rd Edition, Van NostrandReinhold, NY, Chapter-2, Page 33-107

34.Deshmukh, K., Basheer Ahamed, M., Deshmukh, R.R., Khadheer Pasha, S.K., Bhagat, P.R., Chidambaram, K., (2017), Biopolymer Composites With High Dielectric Performance: Interface Engineering, From Book: Biopolymer Composites in Electronics, ISBN 9780128092613, Pages 27-128. https://doi.org/10.1016/B978-0-12-809261-3.00003-6
35.Domenek, S. & Feuilloley, Pierre & Gratraud, Jean & Morel, Marie-Hélène & Guilbert, Stéphane. (2004). Biodegradability of wheat gluten based bioplastics. Chemosphere. 54. 551-9. 10.1016/S0045-6535(03)00760-4.
36.Draget I, Moe T, Skjak-Bræk G and Smidsrød O (2016) Alginates from algae. In: Biopolymer (Hofrichter M and Steinbüchel A, eds.). pp.160-289. Wiley-VCH Verlag GmbH & Co. KGaA, Weinheim.
37.Dybing ST, Smith DE (1991) Relation of chemistry and processing procedures to whey protein functionality: a review. Cultured Dairy Prod J 26:4–12
38.El-Wakil NA, Hassan EA, Abou-Zeid RE, Dufresne A (2015) Development of wheat gluten/ nanocellulose/titanium dioxide nanocomposites for active food packaging. Carbohyd Polym 124:337–346
39.European bioplastics (nova institute) 2019, Applications for bioplastics, Retrieved from URL https://www.european-bioplastics.org/market/applications-sectors/
40.European bioplastics (nova institute) 2019, Bioplastics market data , Retrieved from URL https://www.european-bioplastics.org/market/
41.Fang, Wei & Yang, Sheng & Wang, Xi-Luan & Yuan, Tong-Qi & Sun, Run-Cang. (2017). Manufacture and Application of Lignin-based Carbon Fibers (LCFs) and Lignin-based Carbon Nanofibers (LCNFs). Green Chem. DOI: 10.1039/C6GC03206K.
42.Frollini, E., Silva, C.G. & Ramires E.C. (2013), Phenolic resins as a matrix material in advanced fiber-reinforced polymer (FRP) composites, From book Advanced Fibre-Reinforced Polymer (FRP) Composites for Structural Applications, Chapter 2, Pages 7-43. https://doi.org/10.1533/9780857098641.1.7
43.Gabor, D. & Tita, O. (2012), Biopolymers used in food packaging: a Review, Acta Universitatis Cibiniensis Series E: FOOD TECHNOLOGY Vol. 16(2012), no.2. http://saiapm.ulbsibiu.ro/cercetare/ACTA_E/AUCFT2012_II_3_19.pdf
44.Galego, N., Rozsa, C., Sanchez, R., Fung, J., Vazquez, A. & Tomas, J.S. (2000), Characterization and application of poly (bhydroxyalkanoates) family as composite biomaterials, Polymer Testing- Journal, Volume 19, page 485-492
45.Galus, S. and Kadzinśka, J. (2015). Food applications of emulsion-based edible films and coatings. Trends in Food Science & Technology. Vol. 45 No: 2 Pages: 273-283.
46.Gennadios A, Weller CL (1990) Edible films and coating from wheat and corn proteins. Food Technol 44:63–69
47.Ghanbarzadeh, B., Musavi, M., Oromiehie, A.R, Rezayi, K., Razmi Rad, E., Milani, J. (2007) Effect of plasticizing sugars on water vapor permeability, surface energy and microstructure properties of zein films. LWT 40:1191–1197
48.Giancone, T., Torrieri, E., Masi, P. & Michon, C. (2009), Protein–polysaccharide interactions: Phase behaviour of pectin–soy flour mixture, Food Hydrocolloids, Volume 23, Issue 5, page 1263–1269. https://doi.org/10.1016/j.foodhyd.2008.09.001
49.Glenn, G.M., Orts, W., Imam, S., Chiou, B. & Wood, D.F. (2014),, Starch Plastic Packaging and Agriculture Applications . In Starch Polymers; Elsevier: New York, NY, USA, 2014; pp. 421–452 https://digitalcommons.unl.edu/cgi/viewcontent.cgi?article=2465&context=usdaarsfacpub
50.Goelzer, F.D., Faria Tischer, P., Vitorino, J.C., Sierakowski, M.R & Tischer, C. (2009). Production and characterization of nanospheres of bacterial cellulose from Acetobacter xylinum from processed rice bark, Materials Science and Engineering C 29(2). DOI: 10.1016/j.msec.2008.10.013.

51.Gounga ME, Xu S, Wang Z (2007) Whey protein isolate-based edible films as affected by protein concentration, glycerol ratio and pullulan addition in film formation. J Food Eng 83:521–530
52.Grujic R., Vukic M., Gojkovic V. (2017) Application of Biopolymers in the Food Industry. In: Pellicer E. *et al.* (eds) Advances in Applications of Industrial Biomaterials. Springer, Cham. https://doi.org/10.1007/978-3-319-62767-0_6
53.Guilbert S (1986) Technology and application of edible protective films. In: Food Packaging
54.Guo, M. & Wang, G. (2016), Milk Protein Polymer and Its Application in Environmentally Safe Adhesives. Polymers, Volume 8, DOI: 10.3390/polym8090324.
55.Gutierrez TJ, Morales NJ, Perez E *et al.* (2015) Physico-chemical properties of edible films derived from native and phosphated cush-cush yam and cassavastarches. FoodPackag ShelfLife 3:1–8
56.Gutiérrez, T.J. (2017), Chitosan applications for the food industry , From book Chitosan: Derivatives, Composites and Applications,Chapter8,Page-183-232. doi:10.1002/9781119364849.ch8
57.Hamad W (2006) On the development and applications of cellulosic nanofibrillar and nanocrystalline materials. Can J Chem Eng 84:513–519
58.Hefft, D. (2017), Developments and Properties of Plastic Mimicking Biopolymers for Food Packaging Application, Journal of Applied Packaging research. https://scholarworks.rit.edu/cgi/viewcontent.cgi?article=1095&context=japr
59.Hoffman, A.S. (2002), Hydrogels for biomedical applications, Advanced Drug Delivery Reviews, Volume 54, Issue 1, Page 3–12. DOI: 10.1016/S0169-409X(01)00239-3
60.Hu Y, Liu L, Gu Z, Dan W, Dan N, Yu X (2014) Modification of collagen with a natural derived cross-linker, alginate dialdehyde. Carbohyd Polym 102:324–332
61.Islam, A.M., Phillips, G.O., Sljivo, A., Snowden, M.J. & Williams, P.A. (1997), A review of recent developments on the regulatory, structural and functional aspects of gum Arabic, Food Hydrocolloids, Vol. 11, Issue 4, (October 1997), page. 493-505, ISSN: 0268-005X. https://doi.org/10.1016/S0268-005X(97)80048-3
62.Jahn CE, Selimi DA, Barak JD & Charkowski AO, 2011, The Dickeya dadantii biofi lm matrix consists of cellulose nanofi bres, and is an emergent property dependent upon the type III secretion system and the cellulose synthesis operon, Microbiology, 157, 2733–2744
63.Janjarasskul T., Krochta J.M. 2010. Edible packaging materials. Annual Review of Food Science and Technology 1: 415–448.
64.Jha, S., Rohilla, P. & Singh, K. (2017), Starch Based Packaging Materials: A Review, International Journal of Research and Analytical Reviews, Volume 4, Issue 4, page 209-211, e ISSN 2348-1269. https://ijrar.com/upload_issue/ijrar_issue_523.pdf
65.Jiménez A, Fabra J, Talens P and Chiralt, A (2012). Effect of re-crystallization on tensile, optical and water vapour barrier properties of corn starch films containing fatty acids. Food Hydrocoll., 26: 302-310.
66.Johnson, R.M., Mwaikambo, L.Y. & Tucker, N. (2003), Biopolymers, Chapter 4, Page 15-18, ISBN 1-85957-379-7
67.Kalaivani, R., Nasreen, K. & Vinodhini, V. (2016), An outlook on upshot of bioplastics in India, Paripex-Indian Journal of Research, Volume 5, Page 216-219, DOI : https://www.doi.org/10.36106/paripex
68.Kaplan DL, 1993, Technical overview of biopolymer field, In- Biopolymers: Making Material Nature's Way, F Poillon (ed), Diane Publishing co., Washington DC, pp 19–50
69.Katiyar, V. & Tripathi, N.(2019), Functionalizing gum arabic for adhesive and food packaging applications. DOI: 10.2417/spepro.006897

70.Kenyon, M.M. (1995). Modified starch, maltodextrin, and corn syrup solids as wall materials for food encapsulation, From book Encapsulation and Controlled Release of Food Ingredients, ISBN: 0841231648, American Chemical Society symposium series, Vol. 590, Chapter 4, page 42–50. DOI: 10.1021/bk-1995-0590.ch004

71.Khosravi-Darania, K. and Bucci, D.Z. (2015), Application of Poly(hydroxyalkanoate) In Food Packaging: Improvements by Nanotechnology, Chem. Biochem. Eng. Q., 29 (2), page 275–285, DOI: 10.15255/CABEQ.2014.2260

72.Korma, S.A., Alahmad K., Niazi, S. & Ammar, A.(2016). Production, Application of Chitosan. international journal of research in agricultural sciences. Volume3, Issue 3, ISSN Online: 2348-3997. https://www.researchgate.net/publication/305073232_Production_Application_of_Chitosan

73.Kulkarni, V.S., Shaw, C.(2016), Use of Polymers and Thickeners in Semisolid and Liquid Formulations, Essential Chemistry for Formulators of Semisolid and Liquid Dosages, Academic Press, ISBN 9780128010242, Pages 43-69, https://doi.org/10.1016/B978-0-12-801024-2.00005-4

74.Kurosumi A, Sasaki C, Yamashita Y & Nakamura Y, 2009, Utilization of various fruit juices as carbon source for production of bacterial cellulose by Acetobacter xylinum NBRC 13693, Carbohydrate Polymers, 76, 333–335

75.Liu, I., Kerry. J. F., and Kerry, J. P. (2006). Effect of food ingredients and selected lipids on the physical properties of extruded edible films/casings. International Journal of Food Science and Technology. Vol. 41 No: 3 Pages: 295-302.

76.Luangapai, F., Peanparkdee, M. & Iwamoto, S. (2019). Biopolymer films for food industries: properties, applications, and future aspects based on chitosan, Reviews in Agricultural Science, volume 7, page59-67. doi: https://doi.org/10.7831/ras.7.0_59

77.Luzi, F., Torre, L., Kenny, J. & Puglia, D. (2019), Bio- and Fossil-Based Polymeric Blends and Nanocomposites for Packaging: Structure–Property Relationship, Materials, volume 12. DOI : 10.3390/ma12030471.

78.Mahanti, N.K., Yadav, A., Mangaraj, S., Singh, R. and Arora, S. (2018), Biopolymers as packaging material in food and allied industry. Print ISSN: 2349–8528 Online ISSN: 2321–4902, International Journal of Chemical Studies 2018; 6(2), page 2411-2418 https://www.researchgate.net/publication/325272862

79.Mali S, Grossmann E, Garcí a A, Martino N and Zaritzky E (2004) Barrier, mechanical and optical properties of plasticized yam starch films. Carbohydr. Polym., 56: 129-135.

80.Mallick, P.K. (2010), Thermoplastics and thermoplastic–matrix composites for lightweight automotive structures,From book Materials, Design and Manufacturing for Lightweight Vehicles, Chapter 5, Pages 174-207. https://doi.org/10.1533/9781845697822.1.174

81.Manvi, P.K., Beckers, M., Mohr, B., Seide, G., Gries, T. & Bunge, C.A. (2019), Polymer fiber-based biocomposites for medical sensing applications, from book Materials for Biomedical Engineering Biopolymer Fibers, Chapter 3, Pages 57-88. https://doi.org/10.1016/B978-0-12-816872-1.00003-0

82.Md, Fakruddin. (2012). Biosurfactant: Production and Application. Journal of Petroleum & Environmental Biotechnology. 3. 124. DOI:10.4172/2157-7463.1000124.

83.Meinel, A.J., Kubow, K.E., Klotzsch, E., Fuentes, M., Smith, M.L., Vogel, V., Merkle, H.P. & Meinel, L. (2009), Optimization strategies for electrospun silk fibroin tissue engineering scaffolds, Biomaterials, Volume 30, Page 3058–3067. DOI: 10.1016/j.biomaterials.2009.01.054

84.Meyer M. (2019). Processing of collagen based biomaterials and the resulting materials properties. Biomedical engineering online, 18(1), 24. https://doi.org/10.1186/s12938-019-0647-0

85.Mitelut, A., Tănase, E., Popa, V. & Popa, M. (2015), Sustainable alternative for food packaging: Chitosan biopolymer–A review, AgroLife Scientific Journal, Volume 4, page 52-61, ISSN ONLINE 2286-0126. https://www.researchgate.net/publication/286625311_Sustainable_alternative_for_food_packaging_Chitosan_biopolymer-A_review
86.Mohamed, E.H., Jun, B. & De-Qiang, D. (2019). Biopolymers; Definition, Classification and Applications, Egyptian Journal of Chemistry, Vol. 62, No. 9. pp. 1725 – 1737. DOI: 10.21608/ejchem.2019.6967.1580
87.Mohammad, A., Mohammad, J., Naheed, S., Ramengmawii, Mohammad, N., Mohamed, T. & Hameed, S. (2017), Processing of hybrid polymer composites—a review, From book Hybrid Polymer Composite Materials Processing, Chapter 1, Page 1-22. https://doi.org/10.1016/B978-0-08-100789-1.00001-0
88.Mohammed J.N., Zana Wan Dagang, W.R. (2019) Implications for industrial application of bioflocculant demand alternatives to conventional media: waste as a substitute. Water Sci Technol 15 November; 80 (10): 1807–1822. DOI: https://doi.org/10.2166/wst.2020.025
89.Mohan Bhagyaraj, S., Oluwafemi, O.S., Kalarikkal, N., Thomas, S. & Songca, S. (2016). Biopolymers – Application in Nanoscience and Nanotechnology, From book Recent Advance s in Biopolymers, Chapter 3. http://dx.doi.org/10.5772/62225
90.Mohy Eldin, M.S., Soliman, E., Hashem, A. & Tamer, T.M. (2012), Biopolymer Modifications for Biomedical Applications, from book Infrared Spectroscopy - Life and Biomedical Sciences, InTech, ISBN 978-953-51-0538-1, DOI: 10.5772/35489.
91.Montenegro, M.A., Boiero, M.L., Valle, L. and Borsarelli, C.D. (2012), Gum Arabic: More Than an Edible Emulsifier, From book PRODUCTS AND APPLICATIONS OF BIOPOLYMERS, (ISBN 978-953-51-0226-7), First Edition, Chapter 1, Page 3-26. https://www.intechopen.com/books/products-and-applications-of-biopolymers/gum-arabic-more-than-an-edible-emulsifier
92.Moreira S, Silva NB, Almeida-Lima J, Rocha HAO, Medeiros SRB, Alves Jr C & Gama FM, 2009, BC nanofi bres: in vitro study of genotoxicity and cell proliferation, Toxicology Letters 189, 235–241
93.Morgan JL, Strumillo J & Zimmer J, 2013, Crystallographic snapshot of cellulose synthesis and membrane translocation, Nature, 493, 181–186
94.Narayanaswamy, R., Kanagesan, S., Pandurangan, A., Padmanabhan, P. (2016), Basics to different imaging techniques, different nanobiomaterials for image enhancement, Nanobiomaterials in Medical Imaging, William Andrew Publishing, Pages 101-129, ISBN 9780323417365. https://doi.org/10.1016/B978-0-323-41736-5.00004-2
95.Niaounakis, M. (2015), Introduction, from book Biopolymers: Processing and Products, Chapter 1, Pages 1-77. https://doi.org/10.1016/B978-0-323-26698-7.00001-5
96.Nor Aslan, A.K.H., Mohd Ali, M.D., Morad, N.A. & Tamunaidu, P. (2016), International Conference on Chemical Engineering and Bioprocess Engineering, IOP Conf. Series: Earth and Environmental Science, DOI:10.1088/1755-1315/36/1/012040
97.Onar Çamlıbel, N. (2018). Introductory Chapter: Introduction to "Polyester – Production, Characterization and Innovative Applications", Intech. ISBN: 978-953-51-3881-5. DOI: 10.5772/intechopen.74422.
98.Osman, M.E., Williams, P.A., Menzies, A.R. & Phillips, G.O. (1993), Characterization of Commercial Samples of Gum Arabic, Journal of Agricultural and Food Chemistry, Vol. 41, No. 1, (January 1993), page 71-77, ISSN: 0021-8561. DOI:10.1021/jf00025a016
99.Oz, A.T., Süfer, Ö. & Celebi sezer, Y. (2017), Poly (Lactic Acid) Films in Food Packaging Systems, Food Science & Nutrition Technology (FSNT), Volume 2, Issue 4, ISSN: 2574-2701. https://www.researchgate.net/publication/320711423

100.Ozdemir M, Floros JD (2008) Optimization of edible whey protein films containing preservatives for mechanical and optical properties. J Food Eng 84:116–123

101.Palali, S.N. (2019), Cellulose Nanocrystals: Potential Replacement for Food Packaging. DOI: 10.13140/RG.2.2.23287.57769.

102.Palmer, R.J. (2005). Encyclopedia Of Polymer Science and Technology, Volume 3, ISBN 9780471238966. DOI: 10.1002/0471238961.1612011916011213.a01.pub2.

103.Patachia, S. & Croitoru, C. (2016), Biopolymers for wood preservation , From book Biopolymers and Biotech Admixtures for Eco-Efficient Construction Materials, Chapter 14, Page 305-332. https://doi.org/10.1016/B978-0-08-100214-8.00014-2

104.Patni,, N., Yadava, P., Agarwal, A., & Maroo, V. (2014). An overview on the role of wheat gluten as a viable substitute for biodegradable plastics, Reviews in chemical engineering. 30. 10.1515/revce-2013-0039.

105.Popa, Elisabeta & Popa, Mona & Râpă, Maria & Popa, Ovidiu. (2015). Preparation and characterization of biopolymer blends based on polyvinyl alcohol and starch. Romanian Biotechnological Letters. 20. Page 10306-10315.https://www.researchgate.net/publication/275582605_Preparation_and_characterization_of_biopolymer_blends_based_on_polyvinyl_alcohol_and_starch

106.Popović, S., Peričin, D., Vaštag, Z., Lazić, V., & Popović, L. (2012), Pumpkin oil cake protein isolate films as potential gas barrier coating, Journal of Food Engineering, 110(3), page 374–379, DOI: 10.1016/j.jfoodeng.2011.12.035

107.Radzik, P., Leszczyńska, A. & Pielichowski, K. Modern biopolyamide-based materials: synthesis and modification. Polym. Bull. 77, 501–528 (2020). DOI: https://doi.org/10.1007/s00289-019-02718-x

108.Rai, S., Dutta, P. & Mehrotra, G. (2017), Lignin incorporated antimicrobial chitosan film for food packaging application, Journal of Polymer Materials, volume 34, page 171-183. https://www.researchgate.net/publication/318835566

109.Raman, M. and Gopakuma, K. (2018), Fish Collagen and its Applications in Food and Pharmaceutical Industry: A Review, EC Nutrition 13.12 (2018): 752-767. https://www.ecronicon.com/ecnu/pdf/ECNU-13-00529.pdf

110.Ramos, Ó. L., Pereira, R. N., Silva, S. I., Fernandes, J. C., Franco, M. I., Lopes-da-Silva, J. A., Pintado, M. E., Malcata, F. X. (2012). Evaluation of antimicrobial edible coatings from a whey protein isolate base to improve the shelf life of cheese. Journal of Dairy Science. Vol. 95 No: 11 Pages: 6282-6292

111.Ramos, Ó.L., Silva, S.I., Soares, J.C., Fernandes, J.C., Poças, M.F., Pintado, M.E., & Malcata, F.X. (2012), Features and performance of edible films, obtained from whey protein isolate formulated with antimicrobial compounds, Food Research International, Volume 45, page 351–361. DOI: 10.1016/j.foodres.2011.09.016

112.Rao, M.G., Bharathi, P. & Akila, R.M. (2014), Sci. Revs. Chem. Commun, 4(2), page 61-68, ISSN 2277-2669. https://www.tsijournals.com/articles/a-comprehensive-review-on-biopolymers.pdf

113.Rasal, R.M.; Janorkar, A.V.; Hirt, D.E (2010). Poly (lactic acid) modifications. Prog. Polym. Sci., 35, 338–356.

114.Rhim, J.-W., Hong, S.-I., Ha, C.-S. (2009) Tensile, water vapor barrier and antimicrobial properties of PLA/nanoclay composite films. LWT Food Sci. Technol., 42, 612–617.

115.Robertson GL (2012), Edible, Biobased and Biodegradable Food Packaging Materials, from Book Food Packaging – Principles and Practice, CRC Press Taylor and Francis Group, Page 50-82

116.Rosa, D.S., Lotto, N.T., Lopes, D.R. & Guedes, C.G.F. (2004), Polymer Testing, Volume 23, Issue 1, February 2004, Page 3-8, DOI: http://dx.doi.org/10.1016/S0142-9418(03)00042-4

117.Roy, S.K. & Prabhakar, B. (2010). Tropical Journal of Pharmaceutical Research, February 2010; 9 (1): 91-104. http://www.bioline.org.br/pdf?pr10012

118.Rudin, A. & Choi, P. (2013), Biopolymers ,from book The Elements of Polymer Science & Engineering, Third Edition, Chapter 13, Pages 521-535. https://doi.org/10.1016/B978-0-12-382178-2.00013-4

119.Sahu, M. & Sahoo, P.K. (2017), Bio Polymers: Sustainable Alternative for Food Packaging, National Seminar on Sustainable Future, through Leadership & Technology in collaboration with International Journal of Engineering and Management Research, Page Number: 28-32, ISSN Online: 2250-0758. http://www.ijemr.net/DOC/BioPolymersSustainableAlternativeForFoodPackaging.pdf

120.Said Nurdin, Nor Hidayah Kamin, Malar Vizhli Sivaguru, Nur Syazwani Ghazali, Mohamad Zaki Sahad, Siti Farhana Haron (2017)., Future prospects of biobased detergent derived from Jatropha c. seeds oil (JSO). Aust. J. Basic & Appl. Sci., 11(3): 79-84. http://www.ajbasweb.com/old/ajbas/2017/Special%20issue%20ICCEIB/79-84.pdf

121.Sastri, V. R. (2014), Other Polymers: Styrenics, Silicones, Thermoplastic Elastomers, Biopolymers, and Thermosets, From book Plastics in Medical Devices: Properties, Requirements and Applications (Second Edition), Chapter 9, Pages 215-261. https://doi.org/10.1016/B978-1-4557-3201-2.00009-4

122.Shah, U., Gani, A., Ashwar, B.A., Shah, A., Ahmad, M., Gani, A., Wani, I.A. & Masoodi F.A. (2015), A review of the recent advances in starch as active and nanocomposite packaging films, Cogent Food & Agriculture, 1:1, 1115640, DOI: 10.1080/23311932.2015.1115640

123.Shamala TR, Divyashree MS, Davis R, Kumari KSL, Vijayendra SVN & Raj B, 2009, Production and characterization of bacterial polyhydroxyalkanoate copolymers and evaluation of their blends by Fourier transform infrared spectroscopy and scanning electron microscopy, Indian Journal of Microbiology, 49, 251–258

124.Shankar, S. & Rhim, J.W. (2018). Bionanocomposite Films for Food Packaging Applications. From book Reference Module in Food Science, Elsevier, Page 1-10, DOI: 10.1016/B978-0-08-100596-5.21875-1.

125.Shi, W. & Dumont, M.J.(2014), Processing and physical properties of canola protein isolate-based films, Industrial crops and products, 52, 269-277

126.Shoda M & Sugano Y, 2005, Recent advances in bacterial cellulose production, Biotechnology and Bioprocess Engineering, 10, 1–8

127.Silvipriya KS, Krishna Kumar K, Bhat AR, Dinesh Kumar B, Anish John, Panayappan lakshmanan (2015) Collagen: Animal Sources and Biomedical Application. J App Pharm Sci, 5 (03): 123- 127. https://japsonline.com/admin/php/uploads/1466_pdf.pdf

128.Singh R.S., Saini GK & Kennedy JF, 2008, Pullulan: Microbial sources, production and applications, Carbohydrate Polymers, 73, 515–531

129.Singh, R. & Kaur, N. (2015), Microbial Biopolymers for Edible Film and Coating Applications, From book Advances in Biotechnology, Chapter 12, page 187-216, ISBN 978-93-84588-09-0. https://www.researchgate.net/publication/311576578_Microbial_Biopolymers_for_Edible_Film_and_Coating_Applications

130.Steinbüchel, A. (2006). Non-biodegradable biopolymers from renewable resources: Perspectives and impacts. Current opinion in biotechnology. 16. 607-13. DOI: 10.1016/j.copbio.2005.10.011.

131.Strassberger, Z., Tanase, S., and Rothenberg, G., (2014), The pros and cons of lignin valorisation in an integrated biorefinery, RSC Adv., 2014, 4, 25310–25318. DOI: 10.1039/c4ra04747h

132.Stuchell. Y. M., and Krochta J. M. (1995). Edible Coatings on Frozen King Salmon: Effect of Whey Protein Isolate and Acetylated Monoglycerides on Moisture Loss and Lipid Oxidation. Journal of Food Science. Vol. 60 No: 1 Pages: 28-31.

133.Swain, S.N., Biswal, S.M., Nanda, P.K., Nayak, P.L. (2004), Biodegradable soy-based plastics: opportunities and challenges. J Polym Environ 12(1):35–42

134.Talens, P., Perez-Masia, R., Fabra, M. J., Vargas, M., and Chiralt, A. (2012). Application of edible coatings to partially dehydrated pineapple for use in fruit-cereal products. Journal of Food Engineering. Vol. 112 No: 1-2 Pages: 86-93.

135.Tian, H., Xu, G., Yang, B. & Guo, G. (2011), Microstructure and mechanical properties of soy protein/agar blend films: Effect of composition and processing methods, Journal of Food Engineering, volume 107, Issue1, page 21–26. DOI: 10.1016/j.jfoodeng.2011.06.008.

136.Tripathi, A.D., Srivastava, S.K. & Yadav, A. (2014), Biopolymers Potential Biodegradable packaging material for food industry, From book Polymers for packaging Applications, ISBN 9781926895772. https://www.researchgate.net/publication/311435108

137.Turhan , K. N., Cellulose Based Packaging Films Containing Natural Antimicrobial , Agents, Journal of Hygienic Engineering and Design. https://pdfs.semanticscholar.org/cb79/480adb5fc2742de891334a8926c94b9997f7.pdf

138.U.S. Congress, Office of Technology Assessment (1993), Biopolymers: Making Materials Nature's Way, Washington, DC: U.S. Government Printing Office (September, 1993), ISBN: 0-16-042098-9. www.princeton.edu/~ota/disk1/1993/9313/9313.PDF

139.Vargas, M., Perdones, A., Chiralt, A., Chafer, M., and Gonzalez-Martinez, C. (2011). Effect of homogenization conditions on physicochemical properties of chitosan-based film-forming dispersions and films. Food Hydrocolloids. Vol. 25 No: 5 Pages: 1158-1164.

140.Vasiljevic, L. and Pavlović, S. (2017), Biodegradable Polymers Based on Proteins and Carbohydrates, from book Advances in Applications of Industrial Biomaterials, Springer International Publishing AG, DOI 10.1007/978-3-319-62767-0_5 Page-87-102

141.Velickova, E., Winkelhausen, E., Kuzmanova, S., Alves, V. D., and Moldao-Martins, M. (2013). Impact of Chitosan-Beeswax Edible Coatings on the Quality of Fresh Strawberries (Fragariaananassa cv Camarosa) Under Commercial Storage Conditions. LWT - Food Science and Technology. Vol. 52 No: 2 Pages: 80-92

142.Verbeken, D., Dierckx, S. & Dewettinck, K. (2003), Exudate Gums: Occurrence, production, and applications, Applied Microbiology and Biotechnology, Vol.63, page 10–21, ISSN: 0175-7598. DOI: 10.1007/s00253-003-1354-z.

143.Vroman, I. & Tighzert, L. (2009), Biodegradable Polymers, Materials, Volume 2, Page 307–344. DOI: 10.3390/ma2020307.

144.Vukić, M., Grujić, S. and Odzaković, B.(2017), Application of Edible Films and Coatings in Food Production, Advances in Applications of Industrial Biomaterials, Springer International Publishing Page 121-138, DOI 10.1007/978-3-319-62767-0_7

145.Wang Q, Padua GW (2005) Properties of zein films coated with drying oils. J Agric Food Chem 53:3444–3448

146.Wang, Y., Xiong, Y.L., Rentfrow, K.G., & Newman, M.C. (2013) ,Oxidation promotes cross-linking but impairs film-forming properties of whey proteins, Journal of Food Engineering, Volume 115(1), page 11–19. DOI: 10.1016/j.jfoodeng.2012.09.013

147.Wippermann J, Schumann D, Klemm D, Kosmehl H, Salehi-Gelani S & Wahlers T, 2009, Preliminary results of small arterial substitute performed with a new cylindrical biomaterial composed of bacterial cellulose, European Journal of Vascular and Endovascular Surgery 37, 592–596

148.Wittaya, T., (2012), Protein-Based Edible Films: Characteristics and Improvement of Properties, Structure and Function of Food Engineering, Ayman Amer Eissa, IntechOpen, DOI: 10.5772/48167.

149.Yang, H., Guo, X., Chen, X. and Shu, Z.(2014), Preparation and characterization of collagen food packaging film, J. Chem. Pharm. Res., 6(6):740-745ISSN : 0975-7384. https://pdfs.semanticscholar.org/0b84/9bfac055f1636efb91ff9607d3047c5dca89.pdf

150. Yang, J., Ching, Y.C. & Chuah, C.H. (2019), Applications of Lignocellulosic Fibers and Lignin in Bioplastics: A Review, Polymers, 11, 751. DOI: 10.3390/polym11050751
151. Yanping, Ye. (2018). The development of polyurethane. Materials Science: Materials Review. 1. DOI: 10.18063/msmr.v1i1.507.
152. Yoshinaga F, Tonouchi N & Watanabe K, 1997, Research progress in production of bacterial cellulose by aeration and agitation culture and its application as a new industrial material, Bioscience, Biotechnology and Biochemistry, 61, 219–224
153. Zadeh, E.M., O'Keefe, S. & Kim, Y. (2018), Utilization of Lignin in Biopolymeric Packaging Films, ACS Omega, Volume 3, page 7388-7398. DOI: 10.1021/acsomega.7b01341.
154. Zhao, W., Li, J., Jin, K., Liu, W., Qiu, X. & Li, C. (2015), Fabrication of functional PLGA-based electrospun scaffolds and their applications in biomedical engineering, Materials Science and Engineering, Volume 59, Page 1181–1194. DOI: 10.1016/j.msec.2015.11.026.
155. Zhong C., Cao G., Rong K., Xia Z., Peng T., Chen H., Zhao J. Characterization of a microbial polysaccharide-based bioflocculant and its anti-inflammatory and pro-coagulant activity. Colloids Surf. B iBiointerfaces. 2018;161:636–644. DOI: 10.1016/j.colsurfb.2017.11.042

7

Safety Concerns for Nano Food Packaging Materials

Parvinder Kaur[a], Tanweer Alam[b], Rimpi Foujdar[c], A. A Broadway[a]

[a] *Warner College of Dairy Technology, SHUATS, Allahabad, Uttar Pradesh*
[b] *Director, Government of India, Indian Institute of Packaging, Delhi*
[c] *Department of Food Engineering and Technology, Sant Longowal Institute of Engineering and Technology, Longowal, Punjab*

Abstract

The integration of nanotechnology into food packaging has opened avenues for improved barrier, thermal, mechanical, and even antimicrobial functionality. This technological advancement can overcome problems with conventional packaging, such as low gas barriers and controlled delivery of bioactive compounds. However, the adoption of nanotechnology in food packaging raises a host of complex concerns, including safety, environmental impact, ethical considerations, policy formulation, and regulatory measures. While prevailing expectations align with the concept that the inherent fixed or embedded nature of nanomaterials within a polymer matrix minimizes substantial risks to consumers, the complete toxicity profile of these nanomaterials remains inadequately understood. The toxicity of nanomaterials is not well understood, and concerns about nanoparticle migration, and interactions with food or the environment pose additional safety challenges. However, further research addressing the safety issues of nanotechnologies is needed. Comprehensive research and regulation are needed to ensure the safe use of nanomaterials in food packaging.

Introduction

Food packaging materials development has been driven by the need to protect food products throughout distribution and storage. Nanotechnology offers opportunities to improve technical packaging features and develop sustainable smart packaging and sensors for monitoring product quality over time. However, the inherent permeability of conventional packaging materials to atmospheric gases and water vapour poses challenges in maintaining optimal conditions for food. The integration of nanotechnology into food packaging has emerged as a promising solution to address these challenges. Nanocomposite materials, composed of a polymer matrix and nanoscale fillers, offer a novel approach to enhancing packaging performance while addressing issues related to food spoilage, degradation, and safety.

Nanocomposite materials encompass a range of nanomaterials, including inorganic (i.e., metals and metal oxides), nanoclay, organic nanomaterials, and nanoemulsions (Huang *et al.*, 2018; Jamaleddin *et al.*, 2019; P. Kumar *et al.*, 2020). NMs are used for three types of packaging: improved packaging, active packaging and intelligent packaging, Nanotechnology facilitates the development of improved bio-based packaging, enhancing mechanical strength and barrier performance while maintaining stability under varying environmental conditions (S. Kumar *et al.*, 2020; Zhang *et al.*, 2020). Active packaging harnesses nanostructures to encapsulate bioactive compounds, encompassing antimicrobials, antioxidants, and oxygen scavengers, which can interact with food and extend its shelf life (Motelica *et al.*, 2020). Intelligent packaging releases active compounds in response to environmental triggers such as temperature or pH changes, transforming packaging into an intelligent, actively responsive material (Jamróz *et al.*, 2019).

As the utilization of nanomaterials in food packaging evolves, research endeavors are continuously expanding. Currently, approximately 400 to 500 nanofood packaging products are reportedly available on the market (INRS, 2014). The major companies in the food industry are already conducting research on the use of nanofood packaging technologies, and certain applications are already available on the market. Nano clays and silver nanomaterials are the most commonly used nanomaterials due to their multifaceted benefits (Committee & Scheer, 2023; He & Hwang, 2016; Perera *et al.*, 2023; Rovera *et al.*, 2020; Sarfraz *et al.*, 2021). Nanoclay-based nanocomposites can be manufactured into various forms for food packaging. Companies have developed commercial nanoclay composite products for beverage packaging, bread bags, fresh bags for fruits and vegetables, and plastic trays for packaging chocolate and beer bottles (Perera *et al.*, 2023).

Metals and metal oxide nanoparticles (NPs) have been used to improve barrier capacity and imparting antimicrobial properties in packaging (Bott *et al.*, 2017; Corps Ricardo *et al.*, 2021; Cushen *et al.*, 2014; Du *et al.*, 2018). TiO_2 (titanium dioxide) is added to enhance the barrier properties and stability of the packaging material. PET/TiO_2 composite films are commonly used in polymer-laminated steel cans to package acid beverages (Cui *et al.*, 2023). Silver NPs are used in food packaging to provide antimicrobial and sterilization effects against a broad spectrum of microorganisms. ZnO NPs, also possessing potent antibacterial properties, are used in both food additives and packaging materials. They are particularly effective against gram-positive bacteria (Azizi-Lalabadi *et al.*, 2020). SiO_2 NPs find application in nanotechnology-based food packaging due to their inherent antimicrobial properties that hinder the growth of microorganisms on the packaging surface. However, organic polymer-based nanomaterials such as cellulose, starch, and chitosan are equally utilized in a biodegradable polymeric composite for reinforcement due to their nontoxicity and biodegradability. Chitosan, being a polysaccharide with natural antimicrobial properties, can serve a dual purpose in polymer nano biocomposites both as an active component for its antimicrobial effects and as a reinforcing agent to improve the material's overall properties. Chitosan's role is similar to that of silver or titania nanoparticles in inorganic materials, to impart antimicrobial and reinforcing properties (Gopi *et al.*, 2019; Lu *et al.*, 2019; Montero *et al.*, 2017; Orsuwan & Sothornvit, 2017).

However, there is a potential for the nanomaterials to migrate into food and cause adverse health effects, so the safety assessment of nanomaterials in food packaging requires ongoing vigilance.Migration pathways involve the diffusion, dissolution, and dispersion of nanomaterials into the food matrix (Franz *et al.*, 2020). The potential health implications of nanoparticle ingestion, especially in comparison to larger food ingredients, underscore the importance of understanding migration mechanisms (Ashfaq *et al.*, 2022).

While engineered nanomaterials are typically embedded within the polymer or biopolymer matrix, circumstances such as stress or environmental factors can lead to the desorption and release of nanomaterials into food (Cockburn *et al.*, 2012). Nanomaterials upon ingestion can traverse the gastrointestinal tract and enter human cells, potentially instigating oxidative stress and cellular damage. Acute toxic responses, including inflammation, lesions, and even cancer, have been associated with nanomaterial exposure (Naseer *et al.*, 2018). Moreover, uncontrolled disposal of packaging materials may contribute to environmental contamination and negative impacts on ecosystems (Furtado *et al.*, 2016; Singh *et al.*, 2023).

Risk evaluation requires an in-depth understanding of the characteristics of nanomaterials and their migration behaviour within specific food matrices. Rigorous analytical methodologies are essential to characterizing nanomaterials in various environmental and biological contexts, enabling comprehensive toxicological and ecotoxicological assessments. A systematic approach to assessing the safety of nanomaterials in food packaging is imperative to ensure the benefits of nanotechnology are harnessed responsibly, minimizing risks to both consumers and the ecosystem.

Migration of nanomaterials from the nanocomposite into food

The complexity of nanoparticle migration arises from the interplay of multiple factors that influence the release of nanomaterials from the matrix and their subsequent transfer into food products. The release of nanomaterials can take place through desorption, diffusion, dissolution, and degradation phenomena (Stuparu-Cretu *et al.*, 2023). Similar to traditional polymers, nanomaterials can migrate via Fick's diffusion, which involves the movement of low-molecular-weight constituents from the nanocomposite package into food. Desorption so that spontaneous release of diffuse polymer surface molecules can occur due to weak bonding between matrix and nanomaterials (Jokar *et al.*, 2017). In cases where nanomaterials are chemically reactive, they might dissolve into ionic species within the polymer matrix. These dissolved species can then be released into food, leading to migration (Addo Ntim *et al.*, 2016). The polymer matrix of the nanocomposite can degrade due to various factors like mechanical stress, fatigue, UV exposure, hydrolysis, and interactions with food (Paidari *et al.*, 2021; Perera *et al.*, 2023; Sarfraz *et al.*, 2021).

The nanoparticle migration is influenced by several external factors such as the composition and properties of the polymer or biopolymer matrix, the duration and nature of contact between the nanocomposite and food (e.g., direct contact, indirect contact), and environmental conditions such as temperature and pressure (Noonan *et al.*, 2014). The composition, pH, moisture content, and other characteristics of food can interact with the nanocomposite and affect migration. Foods with high acidity, moisture, or fat content might interact more with the nanocomposite, affecting the migration rate.

Migration assessment on polymer nanocomposites

The safety assessment of nanomaterials requires evaluation of their characteristics, interactions with the matrix, potential for migration, and potential toxicological effects. The migration of nanomaterials in food packaging materials is a critical area of study with implications for food safety. Standardized testing methodologies and thorough analysis are necessary to understand and mitigate the potential risks associated with nanoparticle migration, ensuring the safety of consumers and the environment. Two main types of migration, overall migration (OM) and specific migration (SM), are distinguished. OM quantifies the release of all migrated substances under specific conditions, while SM pertains to the migration of a specific known substance. This migration assessment is a regulatory requirement in EU countries, particularly in the context of food contact materials.

Migration studies are often conducted using food simulants instead of actual food due to the complex nature of food matrices (Metak *et al.*, 2015). Food simulators aim to mimic the physicochemical properties of different food types. These tests replicate the migration that occurs from packaging materials to food, including temperature and testing duration, to reflect real storage conditions. The migration process involves two steps: exposing the packaging material to the food simulant or food, and quantifying transferred migrants in terms of OM or SM.

Numerous studies have explored the migration of nanomaterials in food packaging, most frequently packaging containing nanoclay and nanosilver (Echegoyen & Nerín, 2013; Perera *et al.*, 2023). Nanomaterials approval for food packaging depends on their performance in migration tests (He & Hwang, 2016). Researchers have examined migration from different commercial packaging materials.(Echegoyen *et al.*, 2016) evaluated the transfer of nanoclays from LDPE nanocomposite bags into food simulants i.e. ethanol (10%) and acetic acid (3%) at two test conditions: 40 °C for 10 days and 70 °C for 2 hours. ICP-MS determined the presence of different metals, particularly aluminum, in the food simulants. Maximum migration values observed were 51.56 ng/cm^2 for 'Aisaika' bags, and 24.12 ng/cm^2 for 'Debbie Meyer' bags lower than the legislative migration limit of 10 mg/dm^2. A higher concentration of nanoclay in the composite resulted in higher migration levels. Acetic acid at 3% concentration for 10 days at 40 °C induces the highest migration levels in both bag types.

(Echegoyen & Nerín, 2013) found the migration of silver nanoparticles from commercially available food containers. Migration values are higher when acetic acid (3%) is used as the simulant and when the temperature is set at 40 °C for 10 days. (Song *et al.*, 2011) found similar nanosilver migration from commercial PE bags. (Mackevica & Emil, 2016) also measured nanosilver migration from commercial nanocomposites and reported higher migration into the acidic medium than 10% ethanol. Most of the studies found lower migration in non-acidic aqueous simulants than in acetic acid. It may be due to the higher oxidation rate of elemental metal into ions in acidic media (Paidari *et al.*, 2021). (Metak *et al.,* 2015), investigated migration into several foods versus simulants from commercial PE nano-silver containers and nanosilver-coated PE film. Nano-silver coated film samples showed higher levels of migration, while, nano-silver particles were not detected from nano-silver containers samples. (Addo Ntim *et al.*, 2016) studied the behaviour of silver nanoparticles in different food samples. Their findings revealed the importance of considering both ionic silver and nano-silver particles when assessing potential migration from products that contain nanoparticles. If nanoparticles are detected in a stimulant, it's essential to determine whether they migrated from the packaging or if they are the remains of nano sized particles that dissolved during the migration process (Adeyemi & Fawole, 2023). (Cushen *et al.*, 2014) studied the rate of nano-silver and nano copper migration from the polyethylene matrix into chicken breast. They recognized the level of nanomaterial concentration as the most significant parameter influencing migration compared to exposure time, temperature, and nanoparticle size. (Weiner *et al.*, 2018) studied an inverse relationship between nanoparticle size and mass transfer due to the higher release of smaller nanoparticles compared to larger nanoparticles. The release of silver nanomaterials was primarily due to the oxidation of Ag NPs into Ag+ ions in the presence of aqueous acetic acidic food simulants, indicating the chemical transformation of nanomaterials rather than polymer degradation. Ag_2S and Au NPs exhibited significantly less mass transfer compared to Ag NPs. This emphasized the importance of both nanoparticle stability and composition in designing nanocomposites to minimize potential exposure. (Shi *et al.*, 2021) explored nano copper migration from PP films into food and food simulants. They found that the level of migration depends on the types of food, food simulants, and polymer material structure. Different forms of copper, such as copper ions, crystals, and copper oxide, were found in different simulants. Results have shown that higher temperatures and longer exposure times can lead to increased migration. The migration process involves ion dissolution and matrix degradation, which promote the release of TiO_2 from the film. Diffusion and surface desorption are also possible mechanisms (Cui *et al.*, 2023).

The safety assessment of nanoparticles in food packaging was carried out by the European Food Safety Authority (EFSA) CEP Panel for nano-precipitated $CaCO_3$ (both uncoated and coated) and surface-treated TiO_2. The solubility of the uncoated nano CaCO3 was found to be pH-dependent. The dissolution process was considered rapid enough for full solubilization under conditions mimicking the stomach environment of different population groups (More *et al.*, 2021). In contrast to the non-coated particles, the presence of a fatty acid coating affected the solubilization of the nano particles. Complete and rapid dissolution was not achieved. This led to a nano-specific safety assessment according to EFSA guidelines. The assessment of surface-treated TiO_2 followed the guidelines of the EFSA Scientific Committee. Various tests were conducted to investigate potential migration of surface-treated TiO_2 nanoparticles into food (Gitrowski *et al.*, 2014). The results indicated that the nano particles did not migrated, leading to no exposure risk and no toxicological concerns.

Techniques used for the detection of nanomaterials in food

There are various methods to detect, study, and characterize nanomaterials. Electron microscopy (transmission electron microscopy [TEM] and scanning electron microscopy [SEM]), particle sizing, elemental composition (ICP, sp-ICP-MS, AAS, and EDS), and X-ray diffraction (XRD) are the most common methods.

Electron microscopy techniques (SEM and TEM) are effective for visualizing nanomaterials within a polymer matrix. These techniques are utilized for acquiring information such as aggregation state and dispersion morphology, but low particle concentrations can limit their effectiveness (Noonan *et al.*, 2014). TEM is also not suitable for liquid simulants. Evaporation of the simulant for TEM sample preparation can lead to fragment formation, especially if the nanomaterials change form during drying. The process of solubilizing metallic nanomaterials for analysis might lead to their reduction and re-aggregation during sample preparation, potentially leading to misinterpretation of the results.

ICP-MS determines total elemental concentrations rather than nanomaterial-specific data. It doesn't differentiate between the ionic and particulate forms of nanomaterials. This could be problematic when using inappropriate food simulants or sample preparation methods. While single particle ICP-MS (sp-ICP-MS) is a modification of conventional ICP-MS that aims to detect individual nanomaterials which can distinguish between particulate and ionic species. However, it can be challenging when dealing with complex sample matrices. This limitation can be particularly relevant when analysing redox-sensitive nanomaterials like silver due to the formation of different silver-containing species such as oxides, chlorides, and sulphides. These variations in composition can complicate the identification of species, especially when using techniques like sp-ICP-MS or conventional ICP-MS. Extraction techniques have not been fully developed for sp-ICP-MS analysis, and they provide very little information about the nature of nanomaterial dispersion.

As an alternative approach, the combination of Asymmetric Flow Field-Flow Fractionation (AF4) with Multi-Angle Light Scattering (MALLS) detection and ICP-MS allows for the separation, characterization, and identification of nanomaterials in food simulants (Addo Ntim *et al.*, 2016; Corps Ricardo *et al.*, 2021; Furtado *et al.*, 2016). This approach can face challenges, especially when other matrix components are present in the simulant, potentially affecting unambiguous detection. The combination of AF4/MALLS/ICP-MS enhances characterization by offering both particle size separation and elemental composition identification. This dual-detection approach improves the accuracy of nanoparticle analysis. The success of these techniques relies on using appropriate food simulants that have good dispersion stability and are compatible with both techniques. Thus, the analysis of nanomaterials in food simulants requires careful consideration of the techniques used, as well as their compatibility with the chosen simulants. Combining multiple techniques that offer complementary information can provide a more accurate and comprehensive characterization of nanomaterials.

Nanomaterials route from the nanocomposite into the Human GI tract

The route of nanomaterials from nanocomposites into the human gastrointestinal (GI) tract depends on various factors, such characteristics of the nanomaterials, the composition of the nanocomposite, and the physiological conditions in the GI tract. Nanomaterials might be ingested by humans and potentially absorbed through permeable buccal and sublingual mucosa in the oral cavity, entering the circulatory system through fine capillaries (Sharma et al., 2017). Unabsorbed nanomaterials can pass through the esophagus and reach the stomach, then move into the small intestine. If not absorbed there, they can proceed to the large intestine.

Once ingested, the nanomaterials may eventually come into contact with the GI tract, where they display various behaviours (Hardy et al., 2018). Nanomaterials may interact with the physiological environment in the GI tract. Mucus composition, nanoparticle properties, and electrostatic interactions can influence their interaction. Nanomaterials might undergo changes in size, surface charge, or even aggregation due to these interactions. Depending on their properties, nanomaterials might be absorbed through the cell lining of the GI tract (intestinal epithelial cells) and enter the bloodstream (Naseer et al., 2018). This absorption can occur through various mechanisms, including passive diffusion, endocytosis, and transcytosis. The extent of absorption can vary based on factors such as nanoparticle size, surface properties, and the presence of any surface coatings (Cockburn et al., 2012).

In the small intestine, nanomaterials can adhere to the enteric mucosa and diffuse through it. Absorption can occur through mechanisms like chylomicron uptake by enterocytes. Nanomaterials that are absorbed can be taken up by cells in various organs. The liver and spleen are well-known for significant nanoparticle accumulation (Du et al., 2018).

Nanomaterials, due to their small size and unique properties, might be absorbed intact by the GI tract and could potentially persist in the body. This raises concerns about long-term exposure to these materials. High concentrations of certain nanomaterials could potentially interfere with the normal digestion in the GI tract. This interference might disrupt nutrient absorption and impact overall digestion (Groh & Muncke, 2017). Nanomaterials can be modified by bacteria in the GI tract through secreted enzymes. Additionally, nanomaterials might exhibit antimicrobial activity, which could affect the colonic microbiota and impact overall health. The properties and behaviours of nanomaterials within the GI tract, including their interactions with biomolecules and microbiota, could have negative impacts on human health. Adverse effects might include altered digestion, immune responses, and microbiota disruption.

Nanomaterials immunotoxicity

The immunotoxicity of nanomaterials has been investigated and revealed in various studies. These nanomaterials can impact the human immune system through a range of immune responses, including oxidative stress, inflammatory and anti-inflammatory reactions, as well as genotoxicity. While inflammation is a natural response to external substances and plays a vital role in maintaining human health, uncontrolled inflammation can lead to severe disorders. Nanomaterials have been shown to induce the release of cytokines and chemokines in tissues and organs, such as the spleen and liver, which are crucial for regulating and promoting inflammatory responses (Dainelli et al., 2008).

The immune system's activation of phagocytic cells, including neutrophils, macrophages, and lymphocytes, is a significant aspect of the immune response to nanomaterials exposure. Macrophages can initiate and proliferate an inflammatory response by recognizing, engulfing, and digesting nanomaterials (Du et al., 2018). This interaction often results in the activation of NADPH oxidase, leading to the initiation of reactive oxygen species (ROS) and an oxidative burst. Furthermore, nanomaterials can induce dysfunction in immune cells' organelles, such as the endoplasmic reticulum (ER), mitochondria, and lysosomes. This disruption can result from direct accumulation or indirect subcellular changes (Adeyemi & Fawole, 2023).

ER, a vital organelle responsible for protein synthesis and lipid metabolism, can undergo stress due to nanomaterials exposure, leading to protein misfolding and ER stress. Similarly, interactions with mitochondria can impair mitochondrial function, causing metabolic disorders and reduced cellular energy levels. Nanomaterials can accumulate in lysosomes, causing dysfunction that results in various outcomes, including ROS generation, apoptosis, cytosolic acidification, and necrosis (Cronin et al., 2020).

Additionally, nanomaterials exhibit genotoxicity, which is primarily size-dependent. Smaller nanomaterials tend to have a higher surface area reactivity, leading to increased ROS induction, genotoxic reactions, and DNA damage (Groh & Muncke, 2017). The severity of these effects depends on factors such as concentration, particle size, and physicochemical characteristics of nanomaterials. However, due to the complex interplay of various factors influencing the toxicity of nanomaterials, the precise immunotoxicity mechanism is not fully understood. While the genotoxic effects are primarily size-dependent, the full mechanism of immunotoxicity warrants further investigation through systematic and comprehensive research (More *et al.*, 2021).

Regulations and Guidelines

The regulation of nanomaterials in food packaging is a complex and evolving field that emphasizes risk assessment, exposure evaluation, and migration limits to ensure consumer safety. The approach varies between regions, but both the EU and the US prioritize careful assessment of the safety implications of incorporating nanomaterials into food contact materials(Bumbudsanpharoke & Ko, 2015).

Engineered nanomaterials used in food products fall under the EU's novel food regulation (EU) 2015/2283.Nanomaterials used in food contact materials must be explicitly authorized and undergo a specific risk assessment of their safety(Rauscher et al., 2013).Corrigendum to Commission Directive 90/128/ EEC sets migration limits for plastic-based food packaging. Limits range from 10 mg/dm^2 of packaging material's surface area to 60 mg/kg of food, depending on container size. The European Food Safety Authority (EFSA) provides guidance on risk assessment for nanomaterial applications in the food and feed industry. This includes physicochemical properties, exposure assessment, and hazard characterization (European Commission, 2011).

In the EU, nanomaterials used in food packaging require case-by-case evaluations, even if the bulk substance is approved. Specific physicochemical data and toxicity assessments are needed. In the US, the FDA requires pre-market clearance for nanomaterials as indirect additives in food packaging. The safety evaluation encompasses the overall packaging material.

Only three nanomaterials are approved for use in plastic-packaged foods in the US: silicon dioxide, carbon black, and titanium nitride. Carbon black has a concentration limit of 2.5% w/w in packaged foods. Unapproved nanoparticles require a migration limit of 0.01 mg/kg.

The risk associated with nanomaterials considers both their inherent hazard and potential exposure levels for consumers (More *et al.*, 2021). Exposure occurs through migration from packaging materials or mechanical release from polymer nanocomposites into food. Unlike conventional chemical substances, where defined migration limits or threshold values based on the TTC (Threshold of Toxicological Concern) concept can be applied, the field of nanomaterials presents a challenge. Reliable methodologies are needed to assess nanomaterial exposure accurately, considering the complex properties of nanomaterials. As scientific knowledge and analytical techniques evolve, more comprehensive and standardized methods are expected to develop for accurate risk assessment of nanomaterial exposure in food contact materials.

Implication and Perspectives

The integration of nanomaterials into the food packaging industry holds immense promise for creating improved packaging materials and enhancing food safety by mitigating contamination risks. However, this potential must be carefully weighed against the inherent risks associated with nanomaterials. The existing data gaps surrounding the safety of nanomaterials, including their migration, interactions, toxicity, quantification methods, chronic effects, and environmental impacts, underscore the need for comprehensive research to ensure consumer well-being. Although this investigation is essential, it should not discourage the application of nanomaterials in food packaging; rather, it should guide the formulation of robust regulations based on rigorous migration assessments.

The challenges faced in this domain have sparked a drive for more effective safety assessments, toxicity evaluations, and risk management strategies tailored to nanomaterials in food packaging. Moreover, the evolving understanding of the release, uptake, and toxicity of nanomaterials within food systems necessitates ongoing research efforts. These collective challenges serve as catalysts for informed decision-making within the food packaging industry, fostering a responsible and informed approach to the utilization of nanomaterials.

It is imperative to thoroughly assess the potential adverse health effects arising from exposure to these nanomaterials to ensure the safe utilization of NPs in the food packaging industry. Comprehensive studies and risk assessments are necessary to understand how these NPs interact with the human body. This assessment process will play a pivotal role in guiding regulatory decisions and industry practices to safeguard consumer health and well-being in the face of advancing nanotechnology applications in the realm of food packaging.

References

Addo Ntim, S., Thomas, T. A., & Noonan, G. O. (2016). Influence of aqueous food simulants on potential nanoparticle detection in migration studies involving nanoenabled food-contact substances. Food Additives and Contaminants - Part A Chemistry, Analysis, Control, Exposure and Risk Assessment, 33(5), 905–912. https://doi.org/10.1080/19440049.2016.1174506

Adeyemi, J. O., & Fawole, O. A. (2023). Metal-Based Nanoparticles in Food Packaging and Coating Technologies: A Review. Biomolecules, 13(7), 1092. https://doi.org/10.3390/biom13071092

Ashfaq, A., Khursheed, N., Fatima, S., Anjum, Z., & Younis, K. (2022). Application of nanotechnology in food packaging: Pros and Cons. Journal of Agriculture and Food Research, 7(January), 100270. https://doi.org/10.1016/j.jafr.2022.100270

Azizi-Lalabadi, M., Alizadeh-Sani, M., Divband, B., Ehsani, A., & McClements, D. J. (2020). Nanocomposite films consisting of functional nanoparticles (TiO2 and ZnO) embedded in 4A-Zeolite and mixed polymer matrices (gelatin and polyvinyl alcohol). Food Research International, 137, 109716. https://doi.org/10.1016/j.foodres.2020.109716

Bott, J., Kemmer, D., Franz, R., & St, A. (2017). Trends in Food Science & Technology Critical review of the migration potential of nanoparticles in food contact plastics. 63, 39–50. https://doi.org/10.1016/j.tifs.2017.01.011

Bumbudsanpharoke, N., & Ko, S. (2015). Nano-Food Packaging: An Overview of Market, Migration Research, and Safety Regulations. Journal of Food Science, 80(5), R910–R923. https://doi.org/10.1111/1750-3841.12861

Cockburn, A., Bradford, R., Buck, N., Constable, A., Edwards, G., Haber, B., Hepburn, P., Howlett, J., Kampers, F., Klein, C., Radomski, M., Stamm, H., Wijnhoven, S., & Wildemann, T. (2012). Approaches to the safety assessment of engineered nanomaterials (ENM) in food. Food and Chemical Toxicology, 50(6), 2224–2242. https://doi.org/10.1016/j.fct.2011.12.029

Committee, S., & Scheer, E. R. (2023). Preliminary Opinion on the safety of titanium dioxide in toys. June. https://ec.europa.eu/health/scientific-committees_en

Corps Ricardo, A. I., Avendaño García, S., Guzmán Bernardo, F. J., Ríos, Á., & Rodríguez Martín-Doimeadios, R. C. (2021). Rapid assessment of silver nanoparticle migration from food containers into food simulants using a qualitative method. Food Chemistry, 361(October 2020), 1–5. https://doi.org/10.1016/j.foodchem.2021.130091

Cronin, J. G., Jones, N., Thornton, C. A., Jenkins, G. J. S., Doak, S. H., & Clift, M. J. D. (2020). Nanomaterials and innate immunity: A perspective of the current status in nanosafety. Chemical Research in Toxicology, 33(5), 1061–1073. https://doi.org/10.1021/acs.chemrestox.0c00051

Cui, J., Chen, J., Gu, J., Hong, X., Hong, M., Wu, Z., & Hu, Y. (2023). Migration of TiO2 from PET/TiO2 Composite Films Used for Polymer-Laminated Steel Cans in Acidic Solution. Coatings, 13(5). https://doi.org/10.3390/coatings13050887

Cushen, M., Kerry, J., Morris, M., Cruz-Romero, M., & Cummins, E. (2014). Evaluation and simulation of silver and copper nanoparticle migration from polyethylene nanocomposites to food and an associated exposure assessment. Journal of Agricultural and Food Chemistry, 62(6), 1403–1411. https://doi.org/10.1021/jf404038y

Dainelli, D., Gontard, N., Spyropoulos, D., Zondervan-van den Beuken, E., & Tobback, P. (2008). Active and intelligent food packaging: legal aspects and safety concerns. Trends in Food Science and Technology, 19(SUPPL. 1), S103–S112. https://doi.org/10.1016/j.tifs.2008.09.011

Du, J., Tang, J., Xu, S., Ge, J., Dong, Y., Li, H., & Jin, M. (2018). A review on silver nanoparticles-induced ecotoxicity and the underlying toxicity mechanisms. Regulatory Toxicology and Pharmacology, 98(June), 231–239. https://doi.org/10.1016/j.yrtph.2018.08.003

Echegoyen, Y., & Nerín, C. (2013). Nanoparticle release from nano-silver antimicrobial food containers. Food and Chemical Toxicology, 62, 16–22. https://doi.org/10.1016/j.fct.2013.08.014

Echegoyen, Y., Rodríguez, S., & Nerín, C. (2016). Nanoclay migration from food packaging materials. Food Additives and Contaminants - Part A Chemistry, Analysis, Control, Exposure and Risk Assessment, 33(3), 530–539. https://doi.org/10.1080/19440049.2015.1136844

European Commission. (2011). Commission Regulation (EU) No 10/2011 of 14 January 2011. Official Journal of the European Union, 1–89. https://eur-lex.europa.eu/LexUriServ/LexUriServ.do?uri=OJ:L:2011:012:0001:0089:EN:PDF

Franz, R., Bott, J., & Störmer, A. (2020). Considerations for and guidance to testing and evaluating migration/release of nanoparticles from polymer based nanocomposites. Nanomaterials, 10(6), 1–16. https://doi.org/10.3390/nano10061113

Furtado, L. M., Bundschuh, M., & Metcalfe, C. D. (2016). Monitoring the Fate and Transformation of Silver Nanoparticles in Natural Waters. Bulletin of Environmental Contamination and Toxicology, 97(4), 449–455. https://doi.org/10.1007/s00128-016-1888-2

Gitrowski, C., Al-Jubory, A. R., & Handy, R. D. (2014). Uptake of different crystal structures of TiO2 nanoparticles by Caco-2 intestinal cells. Toxicology Letters, 226(3), 264–276. https://doi.org/10.1016/j.toxlet.2014.02.014

Gopi, S., Amalraj, A., Jude, S., Thomas, S., & Guo, Q. (2019). Bionanocomposite films based on potato, tapioca starch and chitosan reinforced with cellulose nanofiber isolated from turmeric spent. Journal of the Taiwan Institute of Chemical Engineers, 96, 664–671. https://doi.org/10.1016/j.jtice.2019.01.003

Groh, K. J., & Muncke, J. (2017). In Vitro Toxicity Testing of Food Contact Materials: State-of-the-Art and Future Challenges. Comprehensive Reviews in Food Science and Food Safety, 16(5), 1123–1150. https://doi.org/10.1111/1541-4337.12280

Hardy, A., Benford, D., Halldorsson, T., Jeger, M. J., Knutsen, H. K., More, S., Naegeli, H., Noteborn, H., Ockleford, C., Ricci, A., Rychen, G., Schlatter, J. R., Silano, V., Solecki, R., Turck, D., Younes, M., Chaudhry, Q., Cubadda, F., Gott, D., … Mortensen, A. (2018). Guidance on risk assessment of the application of nanoscience and nanotechnologies in the food and feed chain: Part 1, human and animal health. EFSA Journal, 16(7). https://doi.org/10.2903/j.efsa.2018.5327

He, X., & Hwang, H. M. (2016). Nanotechnology in food science: Functionality, applicability, and safety assessment. Journal of Food and Drug Analysis, 24(4), 671–681. https://doi.org/10.1016/j.jfda.2016.06.001

Huang, Y., Mei, L., Chen, X., & Wang, Q. (2018). Recent developments in food packaging based on nanomaterials. Nanomaterials, 8(10), 1–29. https://doi.org/10.3390/nano8100830

INRS. (2014). Manufactured Nanomaterials By 2030. 272.

Jamaleddin, S., Hadi, S., & Pournasir, N. (2019). Properties of active starch-based films incorporating a combination of Ag , ZnO and CuO nanoparticles for potential use in food packaging applications. Food Packaging and Shelf Life, 22(December 2018), 100420. https://doi.org/10.1016/j.fpsl.2019.100420

Jamróz, E., Kulawik, P., & Kopel, P. (2019). The effect of nanofillers on the functional properties of biopolymer-based films: A review. In Polymers (Vol. 11, Issue 4). MDPI AG. https://doi.org/10.3390/polym11040675

Jokar, M., Pedersen, G. A., & Loeschner, K. (2017). Six open questions about the migration of engineered nano-objects from polymer-based food-contact materials: a review. Food Additives and Contaminants - Part A Chemistry, Analysis, Control, Exposure and Risk Assessment, 34(3), 434–450. https://doi.org/10.1080/19440049.2016.1271462

Kumar, P., Mahajan, P., Kaur, R., & Gautam, S. (2020). Nanotechnology and its challenges in the food sector: a review. Materials Today Chemistry, 17, 100332. https://doi.org/10.1016/j.mtchem.2020.100332

Kumar, S., Mudai, A., Roy, B., Basumatary, I. B., Mukherjee, A., & Dutta, J. (2020). Biodegradable hybrid nanocomposite of chitosan/gelatin and green synthesized zinc oxide nanoparticles for food packaging. Foods, 9(9). https://doi.org/10.3390/foods9091143

Lu, H., Ji, N., Li, M., Wang, Y., Xiong, L., Zhou, L., Qiu, L., Bian, X., Sun, C., & Sun, Q. (2019). Preparation of Borax Cross-Linked Starch Nanoparticles for Improvement of Mechanical Properties of Maize Starch Films. Journal of Agricultural and Food Chemistry, 67(10), 2916–2925. https://doi.org/10.1021/acs.jafc.8b06479

Mackevica, A., & Emil, M. (2016). Silver nanoparticle release from commercially available plastic food containers into food simulants. Journal of Nanoparticle Research. https://doi.org/10.1007/s11051-015-3313-x

Metak, A. M., Nabhani, F., & Connolly, S. N. (2015). Migration of engineered nanoparticles from packaging into food products. Lwt, 64(2), 781–787. https://doi.org/10.1016/j.lwt.2015.06.001

Montero, B., Rico, M., Rodríguez-Llamazares, S., Barral, L., & Bouza, R. (2017). Effect of nanocellulose as a filler on biodegradable thermoplastic starch films from tuber, cereal and legume. Carbohydrate Polymers, 157, 1094–1104. https://doi.org/10.1016/j.carbpol.2016.10.073

More, S., Bampidis, V., Benford, D., Bragard, C., Halldorsson, T., Hernández-Jerez, A., Hougaard Bennekou, S., Koutsoumanis, K., Lambré, C., Machera, K., Naegeli, H., Nielsen, S., Schlatter, J., Schrenk, D., Silano, V., Turck, D., Younes, M., Castenmiller, J., Chaudhry, Q., … Schoonjans, R. (2021). Guidance on risk assessment of nanomaterials to be applied in the food and feed chain: human and animal health. EFSA Journal, 19(8). https://doi.org/10.2903/j.efsa.2021.6768

Motelica, L., Ficai, D., Ficai, A., Oprea, O. C., Kaya, D. A., & Andronescu, E. (2020). Biodegradable antimicrobial food packaging: Trends and perspectives. In Foods (Vol. 9, Issue 10). MDPI AG. https://doi.org/10.3390/foods9101438

Naseer, B., Srivastava, G., Qadri, O. S., Faridi, S. A., Islam, R. U., & Younis, K. (2018). Importance and health hazards of nanoparticles used in the food industry. Nanotechnology Reviews, 7(6), 623–641. https://doi.org/10.1515/ntrev-2018-0076

Noonan, G. O., Whelton, A. J., Carlander, D., & Duncan, T. V. (2014). Measurement methods to evaluate engineered nanomaterial release from food contact materials. Comprehensive Reviews in Food Science and Food Safety, 13(4), 679–692. https://doi.org/10.1111/1541-4337.12079

Orsuwan, A., & Sothornvit, R. (2017). Development and characterization of banana flour film incorporated with montmorillonite and banana starch nanoparticles. Carbohydrate Polymers, 174, 235–242. https://doi.org/10.1016/j.carbpol.2017.06.085

Paidari, S., Tahergorabi, R., Anari, E. S., Nafchi, A. M., Zamindar, N., & Goli, M. (2021). Migration of various nanoparticles into food samples; a review. Foods, 10(9), 1–11. https://doi.org/10.3390/foods10092114

Perera, K. Y., Hopkins, M., Jaiswal, A. K., & Jaiswal, S. (2023). Nanoclays - containing bio - based packaging materials : properties , applications , safety , and regulatory issues. Journal of Nanostructure in Chemistry, 0123456789. https://doi.org/10.1007/s40097-023-00525-5

Rauscher, H., Sokull-Klüttgen, B., & Stamm, H. (2013). The European Commission's recommendation on the definition of nanomaterial makes an impact. Nanotoxicology, 7(7), 1195–1197. https://doi.org/10.3109/17435390.2012.724724

Rovera, C., Ghaani, M., & Farris, S. (2020). Nano-inspired oxygen barrier coatings for food packaging applications: An overview. Trends in Food Science and Technology, 97(January), 210–220. https://doi.org/10.1016/j.tifs.2020.01.024

Sarfraz, J., Gulin-sarfraz, T., Nilsen-nygaard, J., & Pettersen, M. K. (2021). Nanocomposites for Food Packaging Applications : An Overview.

Sharma, C., Dhiman, R., Rokana, N., & Panwar, H. (2017). Nanotechnology: An untapped resource for food packaging. Frontiers in Microbiology, 8(SEP). https://doi.org/10.3389/fmicb.2017.01735

Shi, Y. J., Wrona, M., Hu, C. Y., & Nerín, C. (2021). Copper release from nano-copper/polypropylene composite films to food and the forms of copper in food simulants. Innovative Food Science and Emerging Technologies, 67(May 2020), 102581. https://doi.org/10.1016/j.ifset.2020.102581

Singh, R., Dutt, S., Sharma, P., Sundramoorthy, A. K., Dubey, A., Singh, A., & Arya, S. (2023). Future of Nanotechnology in Food Industry: Challenges in Processing, Packaging, and Food Safety. Global Challenges, 7(4). https://doi.org/10.1002/gch2.202200209

Song, H., Li, B., Lin, Q. B., Wu, H. J., & Chen, Y. (2011). Migration of silver from nanosilver-polyethylene composite packaging into food simulants. Food Additives and Contaminants - Part A, 28(12), 1758–1762. https://doi.org/10.1080/19440049.2011.603705

Stuparu-Cretu, M., Braniste, G., Necula, G. A., Stanciu, S., Stoica, D., & Stoica, M. (2023). Metal Oxide Nanoparticles in Food Packaging and Their Influence on Human Health. Foods, 12(9), 1–18. https://doi.org/10.3390/foods12091882

Weiner, R. G., Sharma, A., Xu, H., Gray, P. J., & Duncan, T. V. (2018). Assessment of Mass Transfer from Poly(ethylene) Nanocomposites Containing Noble-Metal Nanoparticles: A Systematic Study of Embedded Particle Stability [Research-article]. ACS Applied Nano Materials, 1(9), 5188–5196. https://doi.org/10.1021/acsanm.8b01241

Zhang, C. wei, Nair, S. S., Chen, H., Yan, N., Farnood, R., & Li, F. yi. (2020). Thermally stable, enhanced water barrier, high strength starch bio-composite reinforced with lignin containing cellulose nanofibrils. Carbohydrate Polymers, 230(July), 115626. https://doi.org/10.1016/j.carbpol.2019.115626

8

FSS Labelling and Display Regulations 2020: A Comprehensive Review for Increased Transparency in the Indian Food Industry

[1]Harish R K and [2]Aastha Bhardwaj

[1]Technical Officer, IIP Bengaluru
[2]Associate Professor, IIP Delhi

Introduction

The landscape of the Indian food industry has undergone a dramatic transformation in recent years. Rising disposable incomes, coupled with the rapid pace of urbanization, have fuelled a surge in demand for convenient and processed food options. Consumers, increasingly pressed for time, are turning towards ready-to-eat meals, packaged snacks, and a wider variety of processed food products. While this trend signifies a growing economy and changing lifestyles, it has also brought to light a critical concern: ensuring food safety and empowering consumers with clear and accessible information. In order to empower consumers to make educated food choices and build trust between brands and consumers, food label transparency is crucial. Transparent labelling is essential for improving public health because it gives precise and unambiguous information about ingredients, nutritional value, and possible allergies. This information helps prevent diseases linked to a certain lifestyle, such diabetes, heart disease, and obesity. Additionally, it guarantees that customers are not misled about the quality or provenance of their food, shielding them from dishonest business tactics. Transparency also promotes market justice by levelling the playing field, making product integrity apparent, and fostering creativity and brand loyalty. Transparent labelling promotes ethical and sustainable consumption while also facilitating international trade by adhering to regulatory norms in many countries.

Recognizing this need for a more transparent and consumer-centric food ecosystem, the Food Safety and Standards Authority of India (FSSAI) introduced the FSS Labelling and Display (L & D) Regulations in 2020 under the Food Safety and Standards Act, 2006 (FSS Act). This landmark legislation marked a significant shift in the way food is labelled and marketed in India. By establishing stricter regulations and mandating the inclusion of comprehensive information on food packaging, the FSSAI aimed to empower consumers to make informed choices while also promoting fair competition within the food industry. This comprehensive review delves into the intricacies of the FSS Labelling and Display Regulations 2020. It explores the key provisions of the regulations, analyzes their impact on various stakeholders, and examines the challenges and opportunities associated with their implementation. By providing an in-depth analysis, this chapter aims to shed light on the significance of these regulations in shaping a more responsible and transparent food industry landscape in India.

The Food Safety and Standards (FSS) Act, 2006: Setting the Stage for Informed Choices

Enacted in 2006, the FSS Act established FSSAI as the central regulatory body responsible for ensuring food safety and promoting a safe and wholesome food supply in India. FSS Act mandates the formulation and enforcement of regulations to ensure food safety across various aspects, including the establishment of standards for food products, specifications for additives and contaminants, and the implementation of stringent hygiene and sanitation practices in food businesses. It also governs the import, manufacture, storage, distribution, sale, and service of food, ensuring that all stages of the food supply chain meet safety standards. Additionally, the Act provides for the establishment of food laboratories for the analysis of food samples, enabling effective monitoring and enforcement of safety regulations. Crucially, the Act empowers the FSSAI to regulate the labelling and display of food products, recognizing the importance of transparency in providing consumers with accurate and clear information. Hence, this commitment to transparency is reflected in the introduction of the FSS Labelling and Display (L&D) Regulations in 2020, which were designed to enhance consumer awareness, protect public health, and ensure that food products in the market are labelled with comprehensive and truthful information

Superseding the Past: From Packaging and Labelling to Labelling and Display

Prior to the FSS L&D Regulations, the Food Safety and Standards (Packaging and Labelling) Regulations, 2011, governed the labelling requirements for pre-packaged food items. However, these regulations primarily focused on aspects like product packaging materials and basic labelling information. The FSS L&D Regulations (2020) represent a significant advancement, shifting the focus from just packaging to a more comprehensive framework encompassing labelling and display requirements. This involves assertion of information such as allergen declarations and nutritional data. The laws also cover e-commerce platforms in response to the expanding online food market. They provide guidelines to guarantee that consumers are provided with accurate and transparent information when making online food purchases. Additionally, the rules provide uniformity in consumer information across all points of sale by promoting transparency not only in food packaging but also in the manner food items are exhibited in restaurants and retail establishments. By superseding the 2011 regulations, the FSS L&D Regulations establish a more robust and transparent labelling framework that empowers consumers to make informed choices about the food they purchase and consume.

FSS Labelling and Display Regulations (2020) vs. Packaging and Labelling Regulations (2011): A Detailed Comparison

The introduction of the FSS Labelling and Display Regulations 2020 marked a significant shift in the way food is labelled and marketed in India compared to the previous Food Safety and Standards (Packaging and Labelling) Regulations, 2011 (Packaging & Labelling Regulations). Here's a detailed comparison highlighting the key differences:

New Inclusions in the FSS Labelling and Display Regulations 2020: A Breakdown with Reasons

The FSS Labelling and Display Regulations 2020 (FSS L&D Regulations) introduced several new inclusions. Here's a detailed breakdown of these inclusions and the potential reasons behind them:

Feature	FSS Labelling and Display Regulations 2020 (FSS L&D)	FSS Packaging and Labelling Regulations 2011	Impact on Consumers and Businesses
Focus	Enhanced consumer information and transparency	Primarily focused on product packaging materials and basic labelling information	Consumers have limited access to detailed information for informed choices. Businesses have less pressure to prioritize transparency.
Mandatory Labelling	Nutritional Labeling (energy, protein, carbohydrates, fat, sodium) - Allergen Labeling (clear and prominent declaration) - List of ingredients (including percentage declaration for specific ingredients) - Net quantity, lot number, expiry date, storage instructions	List of ingredients - Manufacturer information - Net quantity, lot number, expiry date, storage instructions	Consumers gain access to crucial information about nutrient content, potential allergens, and ingredient details. Businesses must prioritize accurate and detailed labelling.
Display Requirements	Specific requirements for font size, color contrast, and information hierarchy for clear and readable labels. - Display regulations for pre-packaged food items at retail stores and restaurants.	Less stringent display requirements focusing primarily on packaging.	Consumers benefit from clear and easily understandable labels. Businesses must ensure proper display of information on both packaging and at points of sale.
E-commerce Compliance	Regulations address online food sales, mandating e-commerce platforms to obtain mandatory labelling information from restaurants and publish it on their websites.	Not addressed	Consumers gain access to crucial information when purchasing food online. E-commerce platforms must comply with labelling regulations.
Front-of-Pack Labelling (FOPNL)	Introduction of the Indian Nutrition Rating (INR) system displayed on FOPNL, providing a quick visual representation of the overall nutritional profile (amendment in 2022).	Not addressed	Empowers consumers with a simplified tool for making informed choices about the nutritional value of food products.

New Inclusions	FSS L&D Regulations Requirement	Reason for Inclusion	Impact on Consumers and Businesses
Mandatory Nutritional Labeling	Declaration of specific nutrients like energy (calories), protein, carbohydrates, fat (including saturated fat), and sodium per serving.	- Empowers consumers to make informed choices based on dietary needs and preferences. - Promotes healthier food choices by highlighting key nutrients.	Consumers gain crucial information to understand the nutritional value of food products. Businesses face pressure to reformulate products or adjust marketing strategies based on their nutritional profile.
Clear and Prominent Allergen Labeling	- Requires clear and easily identifiable declaration of allergens (e.g., peanuts, wheat, milk, etc.) on the label. - Defines allergens and mandates their listing in a specific format.	- Minimizes risks for consumers with allergies by providing clear information about potential allergens in food products. - Promotes transparency and builds trust with consumers.	Consumers with allergies can make safe food choices. Businesses must prioritize accurate allergen information to avoid legal repercussions and potential harm to consumers.
E-commerce Compliance	- Regulations address online food sales. - Mandate e-commerce platforms to obtain mandatory labelling information from restaurants and publish it on their websites.	- Ensures consumers have access to crucial information when purchasing food online. - Creates a level playing field for online and offline food businesses by requiring similar labelling standards.	Consumers can make informed choices about online food purchases. E-commerce platforms must ensure compliance with labelling regulations to avoid penalties.
Specific Display Requirements	- Mandates specific requirements for font size, color contrast, and information hierarchy on labels to ensure clarity and readability. - Establishes display regulations for pre-packaged food items at retail stores and restaurants.	- Improves consumer understanding of information on labels by promoting clear and consistent presentation. - Ensures consistency in how food items are displayed at points of sale, facilitating informed choices.	Consumers benefit from easily understandable labels. Businesses must invest in packaging design and point-of-sale displays that comply with the regulations.
Indian Nutrition Rating (INR) System (introduced through amendment in 2022)	- Requires the display of the INR logo on the Front-of-Pack Label (FOPNL). - The INR system assigns a star rating (½ star to 5 stars) based on the overall nutritional profile of the food product.	- Provides a simplified visual tool for consumers to quickly assess the nutritional value of food products. - Potentially incentivizes businesses to reformulate products to achieve higher INR ratings, promoting healthier options.	Consumers gain a quick and easy way to compare the nutritional value of different products. Businesses may face pressure to improve the nutritional profile of their products to remain competitive.

The FSS Labelling and Display Regulations 2020 (FSS L&D Regulations) are a comprehensive set of guidelines governing the labelling and display of pre-packaged food items in India. To ensure clear understanding, these regulations are divided into various sections, each addressing specific aspects. Here's a detailed breakdown of the key clauses within each section:

Part I: Preliminary

This section lays the groundwork by defining essential terms used throughout the regulations.

- **Clause 1.1 (Short Title):** Introduces the regulations as the "Food Safety and Standards (Labelling and Display) Regulations, 2020."
- **Clause 1.2 (Commencement):** Specifies the date on which the regulations came into effect.

Part II: General Labelling Requirements

This section outlines the mandatory information that must be displayed on the label of every pre-packaged food item.

- **Clause 2.1 (Declaration of Food):**
    - Requires the label to prominently display the name of the food, accurately reflecting its true nature.
    - Imagine a sample label with the product name prominently displayed at the top, such as "Kissan Tomato Ketchup."
- **Clause 2.2 (List of Ingredients):**
    - Mandates a complete list of ingredients in descending order of their composition by weight or volume at the time of manufacturing.
    - Picture a label with a section titled "Ingredients" followed by a list of ingredients starting with the one present in the highest quantity and going down to the least. For example, "Tomatoes (60%), Sugar (20%), Vinegar (15%), Salt (3%), Spices (2%)"
- **Clause 2.3 (Net Quantity):**
    - Specifies the declaration of the net quantity of the food in the container, expressed in the appropriate metric units (grams, milliliters, liters etc.).
    - Imagine a label with a clear statement mentioning the net quantity of the product. For instance, "Net Content: 500g"
- **Clause 2.4 (Lot/Batch Identification):**
    - Requires identification of the lot or batch of food by a code, ensuring traceability for quality control purposes.

    - Picture a label with a section mentioning "Lot No." or "Batch No." followed by a unique alphanumeric code.
- **Clause 2.5 (Date of Manufacture/Packing):**
    - Mandates the declaration of the date of manufacture or packing in a clear and unambiguous format (e.g., DD/MM/YYYY).
    - Imagine a label with a section labeled "Mfg Date" or "Packing Date" followed by the date of manufacturing or packing.
- **Clause 2.6 (Expiry Date):**
    - Requires the declaration of the expiry date by which the food should be consumed to ensure safety and quality.
    - Picture a label with a section labeled "Exp. Date" followed by the expiry date.
- **Clause 2.7 (Instructions for Use):**
    - Specifies that the label must provide clear instructions for storage and use of the food product to maintain its quality and safety.
    - Imagine a label with a section titled "Instructions" or "Storage and Use" that provides clear directions on how to store and use the product for optimal quality and safety.
- **Clause 2.8 (Name and Complete Address of Manufacturer/Distributor/Importer):**
    - Requires the label to display the name and complete address of the manufacturer, packer, distributor, or importer of the food product.
    - Picture a label with a section mentioning "Manufactured by" or "Distributed by" followed by the complete name and address of the responsible entity.
- **Clause 2.9 (Customer Care Contact Information):**
    - Encourages, but doesn't mandate, the inclusion of customer care contact information for further inquiries or feedback.
    - Imagine a label with a phone number or email address for customer care, though this is not mandatory.
- **Clause 2.10 (Declaration of Non-Vegetarian Food):**
    - Requires clear and prominent declaration if the food contains non-vegetarian ingredients (e.g., meat, eggs).
    - Imagine a label with a clear vegetarian or non-vegetarian symbol, or a text statement specifying if the

Part III: Display Requirements

This section focuses on how the mandatory information on the label should be presented for optimal clarity and readability.

- **Clause 3.1 (General Requirements):**
  - Emphasizes clear and contrasting colors for the label background and text to ensure readability.
  - Specifies minimum font size based on the size of the container to ensure legibility.
  - Imagine a label with a background color that contrasts well with the text color, making the information easy to read. The font size should be large enough to be easily readable, especially on smaller packages.
- **Clause 3.2 (Language):**
  - Mandates that the particulars of declaration must be in English or Hindi in Devnagri script.
  - Imagine a label with the mandatory information displayed in either English or Hindi (Devanagari script).
- **Clause 3.3 (Location and Presentation):**
  - Specifies that the label should be placed on the principal display panel (PDP) of the container for easy visibility.
  - The principal display panel (PDP) is typically the front of the package where most of the important information is displayed. Imagine the label with the required information prominently displayed on the front of the packaging.
- **Clause 3.4 (Group Declaration):**
  - Allows grouping related information (e.g., nutritional information) under a common heading for better organization.
  - Imagine a section on the label titled "Nutritional Information" where all details related to nutrients are grouped together.

Part IV: Labelling of Pre-packaged Foods

This section addresses specific labelling requirements for various categories of pre-packaged food items.

- **Clause 4.1 (Irradiated Food):**
  - Requires irradiated food to be labelled with the specific radiation type and purpose of irradiation.
  - Imagine a label for irradiated food with a statement mentioning "Irradiated" followed by the type of radiation used and the purpose of irradiation.

- **Clause 4.2 (Genetically Modified (GM) Food):**
  - Mandates clear and prominent labelling of GM food products, adhering to specific regulations.
  - Imagine a label for GM food with a clear statement like "Contains Genetically Modified Organisms" or "GM Soyabean Oil" following specific FSSAI guidelines.
- **Clause 4.3 (Food Additives):**
  - Requires a complete list of food additives present in the food, along with their individual functional class names.
  - Imagine a label with a section titled "Food Additives" that lists all the additives used in the product, along with their functional class names (e.g., acidity regulator, emulsifier).
- **Clause 4.4 (Added Sugars, Saturated Fatty Acids and Trans Fatty Acids):**
  - Mandates the declaration of added sugars, saturated fatty acids, and trans fatty acids content per serving.
  - Imagine a label with a nutritional information panel that specifically mentions the amount of added sugars, saturated fatty acids, and trans fatty acids per serving size.
- **Clause 4.5 (Nutritional Information):**
  - Specifies the format and mandatory declaration of nutritional information per serving size, including energy, protein, carbohydrates (including sugars), fat (including saturated fat, trans fat), and sodium.
  - Imagine a dedicated nutritional information panel on the label that presents details like energy, protein, carbohydrates (including sugars), fat (including saturated fat, trans fat), and sodium per serving size in a standardized format.
- **Clause 4.6 (Claims):**
  - Regulates the use of claims on food labels, ensuring they are truthful, verifiable, and not misleading for consumers.
  - Imagine labels with claims like "high in protein" or "low in fat" that are truthful, based on scientific evidence, and not misleading consumers about the product's nutritional value.
- **Clause 4.7 (Vegetarian and Non-Vegetarian Food):**
  - Elaborates on the labelling requirements for vegetarian and non-vegetarian food products, including symbols for easy identification.

- Imagine labels for vegetarian products displaying a green vegetarian symbol as specified by the FSSAI. Non-vegetarian products may display a non-vegetarian symbol or a clear text statement specifying the presence of non-vegetarian ingredients.

Part V: Labelling of Imported Food

This section outlines specific labelling requirements for imported food items:

- **Clause 5.1 (General Requirements):**
  - Mandates that imported food products comply with all the general labelling requirements mentioned in Part II

Part VI: Exemptions

This section outlines specific exemptions from certain labelling requirements for certain categories of food items:

- **Clause 6.1 (Exemptions):**
- Provides exemptions for small packages (less than 12 cm2 label area), fresh fruits and vegetables, unbranded pre-packaged foods, and certain traditionally consumed food items under specific conditions.
- Imagine small packages of candies or nuts that might be exempt from displaying the complete list of ingredients if the label area is less than 12 cm2. However, they would still need to comply with other essential labelling requirements like the name of the food and expiry date. Fresh fruits and vegetables like apples or bananas typically wouldn't require labels due to this exemption. Unbranded pre-packaged foods sold loose, like unpackaged cookies at a bakery, might also be exempt from some labelling requirements. Traditionally consumed items like homemade pickles or sweets prepared for local sale might have relaxed labelling requirements based on specific FSSAI guidelines.

Part VII: Enforcement

This section highlights the enforcement mechanisms for ensuring compliance with the regulations:

- **Clause 7.1 (Enforcement Authority):**
  - Designates the Food Safety and Standards Authority of India (FSSAI) as the enforcement authority.
  - Imagine the FSSAI as the government body responsible for overseeing the implementation and enforcement of these labelling regulations.
- **Clause 7.2 (Power to Enter and Inspect):**
  - Empowers designated officers to enter and inspect premises where food is manufactured, processed, stored, or sold to verify compliance.

- Authorized FSSAI officials can visit food manufacturing units, processing facilities, storage warehouses, and retail stores to inspect labelling practices and ensure adherence to the regulations.

- **Clause 7.3 (Penalties):**
  - Outlines penalties for non-compliance with the regulations, including confiscation of food products and imposition of fines.
  - In case of violations, the FSSAI can take action, which may involve confiscating food items that don't comply with labelling regulations and imposing fines on the responsible parties.

Part VIII: Repeal and Saving

This section concludes the regulations:

- **Clause 8.1 (Repeal):**
  - Repeals the previous Food Safety and Standards (Packaging and Labelling) Regulations, 2011.
  - The FSS L&D Regulations supersede the previous labelling regulations, effectively cancelling them.
- **Clause 8.2 (Saving):**
  - Clarifies that anything done or any action taken under the repealed regulations shall be deemed to have been done or taken under the FSS L&D Regulations.
  - Any actions or approvals related to labelling under the previous regulations remain valid under the new FSS L&D Regulations.

Breakdown of Amendments to FSS Labelling and Display Regulations (2020)

Based on the provided links, here's a detailed breakdown of the three amendments you requested:

Amendment 1: 10th September 2021 (Gazette Notification_Oil_Sweetener_16_09_2021.pdf)

This amendment focuses on regulating the labelling of sweeteners and specific categories of oils. Here's a summary of the key points:

- **Sweeteners:**
  - Requires a declaration of all sweeteners used in the food product, including both natural and artificial sweeteners.
  - Specifies that the quantity of each sweetener needs to be declared per serving.

    - This amendment aims to provide consumers with more transparent information about the types and amounts of sweeteners present in their food.
- **Vegetable Oils**
    - Mandates the declaration of the specific type of vegetable oil used in the product (e.g., soybean oil, sunflower oil, etc.).
    - Previously, labelling vegetable oil generically was permissible.
    - This amendment enhances transparency for consumers who may have specific preferences or allergies related to different vegetable oils.

Amendment 2: 09th September 2022 (Gazette Notification_FSS_Label_12_09_2022.pdf)

This amendment addresses labelling requirements for specific food categories and clarifies certain labelling practices. Key points include:

- **Chewing Gum:**
    - Requires labelling on chewing gum to include instructions for proper disposal, discouraging littering.
    - This amendment promotes responsible waste management practices.
- **Principal Display Panel (PDP):**
    - Provides a clearer definition of the PDP, emphasizing its significance as the primary surface for displaying essential information.
    - This clarification helps ensure that crucial information is prominently displayed for consumer visibility.
- **Other Labelling Clarifications:**
    - The amendment addresses labelling requirements for products sold as a single unit within a multi-packaged product.
    - It also clarifies the labelling of food additives and combination food products.

Amendment 3: 11th October 2022 (Gazette Notification_Labelling_14_10_2022.pdf)

This amendment specifically focuses on labelling regulations for pan masala products. Here's the key takeaway:

- **Pan Masala Labelling:**
    - Mandates clear and prominent declaration on the label that pan masala is "injurious to health."

- This disclaimers aims to raise awareness about the potential health risks associated with consuming pan masala.

Note:

- These regulations are subject to amendments as deemed necessary by the FSSAI.
- The FSSAI website (https://www.fssai.gov.in/) provides access to the official document containing the complete and up-to-date regulations for reference.

Impact of FSS Labelling and Display Regulations on the Food Industry in India

The FSS Labelling and Display Regulations, 2020 (FSS L&D Regulations) have significantly impacted the Indian food industry, influencing various aspects of food production, packaging, and marketing. Here's a detailed breakdown of these impacts:

(a) Enhanced Transparency and Consumer Awareness:

- **Detailed Labelling:** The regulations mandate comprehensive labelling with information like nutritional content, ingredients, allergens, and expiry dates. This empowers consumers to make informed choices based on their dietary needs and preferences.
- **Front-of-Pack Labelling (FoPL):** The introduction of the INR system (star rating based on nutritional profile) on the FoPL simplifies nutritional information for consumers, allowing them to quickly assess a product's healthfulness.

(b) Standardized Practices and Increased Compliance Costs:

- **Uniform Labelling Standards:** The FSS L&D Regulations establish uniform labelling standards across the country, streamlining compliance procedures for food businesses. However, this may require some businesses to adapt their labelling practices to meet the new requirements.
- **Compliance Costs:** Implementing the new labelling requirements might involve investments in packaging redesign, label printing, and potentially reformulating products to achieve better INR ratings.

(c) Improved Food Quality and Potential for Innovation:

- **Focus on Healthier Options:** The emphasis on nutritional information and the INR system can incentivize food businesses to develop or reformulate products with healthier ingredients and lower levels of fat, sugar, and sodium.

- **Innovation in Packaging Design:** To accommodate the new labelling requirements and enhance product appeal, the regulations might encourage innovation in packaging design and information presentation.

(d) Challenges and Areas for Consideration:

- **Compliance Burden for Small Businesses:** The cost of adhering to the regulations might pose a bigger challenge for smaller food businesses compared to larger corporations.
- **Understanding Complex Regulations:** The intricacies of the FSS L&D Regulations, especially for amendments, might require educational initiatives to ensure clear understanding among food businesses.
- **Effective Enforcement:** Robust enforcement mechanisms are crucial to ensure all food businesses comply with the regulations and maintain a level playing field in the market.

Indian Nutrition Rating (INR) and FOPNL

A Closer Look the Indian food industry has witnessed a significant shift with the introduction of the **Indian Nutrition Rating (INR) System** and the **Front-of-Pack Labelling (FoPL)** system. Let's delve deeper into these aspects:

(a) Indian Nutrition Rating (INR) System:

The Introduction of INR system assigns a star rating from ½ star to 5 stars to pre-packaged food products based on their overall nutritional profile, represents a significant step toward simplifying nutrition information for consumers. By considering factors such as calories, saturated fats, sugars, sodium, protein, and dietary fiber content, the system provides a quick and easy way for consumers to assess the healthiness of a product at a glance, making it particularly valuable for those who may not have the time or expertise to analyze detailed nutritional information panels. This system promotes informed choices by allowing consumers to compare INR ratings and select products that align with their dietary needs and goals. Additionally, the INR system incentivizes food manufacturers to reformulate products or introduce healthier options to achieve better ratings, catering to the growing demand for health-conscious food choices. However, challenges such as the potential oversimplification of nutrition, the influence of marketing strategies focused solely on improving INR ratings, and the limited scope of the system, which is not yet mandatory for all food categories, highlight the need for careful consideration.

(b) Front-of-Pack Labelling (FoPL) System:

The FoPL system mandates the display of essential information, such as the INR logo, salt (sodium) content, and fat content, prominently on the front of food packages in a clear, consistent, and easily understandable format. This approach significantly increases consumer awareness by ensuring that crucial nutritional information is readily visible, encouraging shoppers to consider a product's nutritional value before making a purchase. The standardized presentation of FoPL across different brands facilitates easier comparison of products within the same category, aligning India with global practices that promote healthy eating habits. However, challenges remain, such as the limited space for displaying all details found in comprehensive nutritional panels and potential variations in design and placement across brands, which can affect clarity.

Conclusion

In conclusion, the FSS Labelling and Display Regulations (2020), alongside the Indian Nutrition Rating system, represent a transformative step towards a more transparent and consumer-focused food industry in India. By empowering consumers with essential information, promoting transparency, and encouraging healthier food options, these regulations facilitate informed dietary choices that benefit individual well-being and foster a responsible food ecosystem. Furthermore, they present a valuable opportunity for the food industry to innovate and adapt, ensuring it meets the evolving preferences of consumers while adhering to regulatory standards. The long-term success of these initiatives hinges on continued collaboration among stakeholders, ongoing monitoring of their impact, and the flexibility to incorporate new technologies and trends, ultimately paving the way for a healthier future for all.

Reference

https://www.fssai.gov.in/upload/uploadfiles/files/Compendium_Labelling_Display_23_09_2021.pdf

9

Regulatory Aspects of Nano-Packaging and Food Contact Materials

P. Narender Raju and Deepak Gujjar

Food Packaging Laboratory, Dairy Technology Division, ICAR-National Dairy Research Institute, Karnal, Haryana

Introduction

Nanotechnology is a rapidly expanding branch of science that has a critical role to play in raising the standard of living for people in the modern world. The global market for nanotechnology in food packaging is valued at 20 billion in 2023 and is expected to grow to 46.50 billion by 2031, with a compound annual growth rate (CAGR) of 11.32% from 2024 to 2031 (InsightAce Analytic, 2023). The rapid expansion of nanotechnology has paved the way for number of industrial areas, including food processing, food contact materials, and agricultural output (Mohammadi & Jafari, 2021). Nanotechnology is the fastest-growing field of science that significantly enhances the quality of life in today's world. This rapid advancement has opened up numerous industrial sectors, such as food processing, food contact materials, and agricultural production. A natural advancement is the integration of nanotechnology into food packaging. This includes enhanced packaging, active packaging, and a new category known as smart or intelligent packaging. Food contact materials, such as antimicrobial metals and oxides, biodegradable nanocomposites, and nanosensors, are utilized to create active and intelligent packaging. These materials enhance thermal and mechanical durability, increase safety and extend the shelf life of food, and facilitate the monitoring of packaged food (Kah *et al.*, 2013). At the nanoscale, material properties change significantly, making it challenging to predict, identify, evaluate, and manage the health, safety, and environmental risks associated with nanomaterials. For example, nanoparticles (NPs) of 50 nm can enter cells, those of 70 nm can penetrate the lungs, and 30 nm NPs can cross blood-brain barriers. These particles can travel to various tissues and organs, including the brain, bloodstream, lymph nodes, and nervous system, potentially causing toxicity, intestinal diseases, and cancer in different parts of the body (Rieux *et al.*, 2006). Globally, efforts are underway to regulate the safe production, utilization, and handling of nanomaterials. These efforts involve implementing legislation, regulations, as well as offering non-binding recommendations and guidelines. The US Food and Drug Administration (FDA) publicly introduced the concept of nanotechnology and its application in the production of nanoproducts for the first time (Anandharamakrishnan, 2014). Several conferences and symposia have taken place in Asia and Oceania, aiming to elucidate and share information regarding the risks of nanotechnology in the food industry up to now. Because of the scarce knowledge about the dangers associated with nanomaterials, these extremely small-scale materials have garnered significant concern regarding health risks. Failing to introduce nanotechnology-based food products into the market could potentially undermine the advantages of nanotechnology in the food industry unless there is a clear definition, public discussions on food safety evaluations, and appropriate regulatory measures for nutrition (Peters *et al.*, 2014). In reference to the aforementioned issue, this chapter provides insights into the various regulations governing the incorporation of nanoparticles in food contact materials across different countries.

Nanoparticles

Definition of nanoparticles

Nanoscience is a scientific field that investigates the characteristics of matter at the nanoscale, emphasizing the distinctive properties of solid-state materials that arise due to their size (Mulvaney, 2015). Nanotechnology encompasses the synthesis, engineering, and utilization of materials falling within the size range of 1 to 100 nanometers, referred to as nanoparticles (Hasan, 2015). These particles consist of only a few hundred atoms, possessing unique properties distinct from larger quantities of the same material, attributed to their diminutive size and high surface area-to-volume ratio.

Classification of nanoparticles (NPs)

NPs can be of different shapes, sizes, and structures. They can be spherical, cylindrical, conical, tubular, hollow core, spiral, etc., or irregular (Ealia & Saravanakumar, 2017).

NPs classification based on composition

1. **Organic NPs**

   This category includes nanoparticles composed of proteins, carbohydrates, lipids, polymers, or any other organic compound (Pan & Zhong, 2016). Examples that stand out in this category include dendrimers, liposomes, micelles, and protein complexes such as ferritin. Today, organic NPs are mostly used in the biomedical field in targeted drug delivery and cancer therapy.

2. **Carbon-based NPs**

   This type of nanoparticle is mainly made of carbon atoms (Ealia & Saravanakumar, 2017). Some examples of this category are carbon black NPs, carbon quantum dots, and fullerenes. These types of nanoparticles are mainly used in bioimaging, photovoltaic devices, and energy storage.

3. **Inorganic NPs**

   This class of nanoparticles is not made of organic materials and carbon. Some of the examples of inorganic materials used are metal, ceramic, and semiconductor. These kinds of nanoparticles are used in food packaging material to increase their thermal, mechanical and gas barrier properties of materials (Ashfaq *et al.*, 2022).

Application of nanomaterials in food Contact materials

1. **Packaging Nanomaterials for beverage packaging**: Nanoparticles are used in food contact material to increase the barrier resistance of the packaging material to increase the shelf life of food products. The Cornelius Group, a division of Advanced Nanotechnology, has developed a zinc oxide nanoparticle known as NanoZ, which serves as a protective barrier against ultraviolet (UV) light. NanoZ can be utilized both internally within plastic materials and as a coating to prevent UV light penetration. One of NanoZ's key benefits is its capability to effectively block a broad spectrum of UV light while maintaining high levels of transparency. (www.cornelius.co.uk)
2. **Barcoding of food packaging**: Barcoding technology applied to food packaging utilizes advanced techniques like Quantum dots for enhanced authentication. These barcodes serve as a reliable means to verify the authenticity and traceability of packaged food products. Ingenia Technology Ltd. has developed the Laser Surface Authentication (LSA) system, which utilizes quantum dots to identify unique patterns present in materials like paper, plastic, metal, and ceramics. According to the available documentation, this technology is capable of authenticating a variety of items including papers, plastics, metals, credit cards, and packaging (Cowburn, 2008).
3. **Nanoscale freshness indicator**: Freshness indicators offer real-time feedback on food quality, contributing to enhanced consumer confidence and reduced food waste. Active packaging is the fastest growing segment

in food packaging sector. Crown Bio-Technology Ltd. has unique ideas in applications close to the nano market, related to the release of various elements, compounds that create freshness (use in packaging), and durable packaging (recyclable) (www.crownbio.com).

4. **Nanocomposites with a long shelf life**: Nanova has developed a polyethylene terephthalate (PET) nanocomposite with one year's shelf life. The company was looking to take advantage of Nanotalc's unique features. Nanotalc composite is mineral-organic at the molecular level. In this nanocomposite, some nanocrystals of magnesium silicate are chemically attached to the organic polymer to form a huge molecule. As a result, the produced polymer composite is transparent and colorless, even with 20%. The addition of magnesium nanosilicate also improves the polymer's blocking properties against all gases and increases its thermal stability by about 100°C (DePolo & Baird, 2009).
5. **Gas barrier coatings**: Nanoparticles are added to food contact materials to enhance the cross-linking of the material, thereby improving its barrier properties. In context to above, InMat fabricates gas-blocking polymer coatings for packaging applications, which are environmental-friendly. The level of blockage created by these coatings is the same on different substrates. Production of Nanolok nano-coatings begins with an aqueous suspension of nano-dispersed silicates in a polymer substrate (www.inmat.com).
6. **Antibacterial plastic coatings**: PChem developed antibacterial plastic coatings for food packaging, incorporating silver NPs. The company claims that it is possible to use nanosilver as an antimicrobial coating and light reducer in packaging, films, and storage containers (www. nanopchem. com).
7. **Flexible blocking material**: ConverTec produces flexible barrier materials for pharmaceutical, medical, and food packaging utilizing corona beam technology, enabling the detection of minute imperfections such as pores, bumps, and formulation defects measuring less than 0.5 nm in size (www.coronabeam.com).

Risk associated with nanoparticles in packaging and food material

The incorporation of nanoparticles in food contact materials offers several benefits, such as increased barrier and mechanical properties. However, despite these benefits, there are also certain risk issues associated with nanoparticles. The toxicity related to the use of nanoparticles is mentioned below:

Genotoxicity

Nanomaterials, due to their high surface area-to-volume ratio, can be easily absorbed via the gastrointestinal route and come into direct contact with cellular proteins, cell material, and DNA. The genotoxicity caused by nanoparticles involves either direct contact with genetic substances (RNA and DNA) or indirect contact through reactive oxygen species (ROS) (Barnes *et al.*, 2008). Both in vivo and in vitro studies have shown the formation of ROS by nanoparticles (Khan *et al.*, 2012).

Carcinogenicity

Exposure to nanoparticles may cause mutations and DNA damage, potentially leading to the generation of carcinogenic substances. The generation of reactive oxygen species (ROS) can result in aging, carcinogenesis, and other diseases due to tissue degradation (Luo *et al.*, 2011). Nanoparticles are highly reactive due to their outer shell electrons, which interact with proteins and initiate various toxic reactions, ultimately leading to inflammatory responses (Bergamaschi *et al.*, 2006).

Regulation of different countries on the incorporation of nanoparticles in food contact materials

There are several applications of nanoparticles in food contact material as mentioned above. But before the application of these nanoparticles in food contact materials their risk factor and safety should be considered by different stakeholders and manufacturers. To address these concerns, regulatory bodies around the world have developed guidelines and regulations to ensure the safe use of nanotechnology in Food contact material. Regulations given by different different countries are mentioned below:

European Food Safety Authority (EFSA)

Nanomaterial in food contact material (FCM)

According to Regulation (EC) No 1935/2004, food contact materials (FCM) must not pose a risk to human health, must not cause any unacceptable changes in food composition, and must not compromise the sensory properties of the food. However, this regulation does not specifically address nanomaterials.

Regulation (EU) 10/2011 sets out safety requirements for plastic materials and articles intended to come into contact with food. This regulation is a specific measure for plastic food contact materials as mentioned in the European Framework Regulation (EU) 1935/2004. This regulation has given list of substances authorized for use in plastic and their migration limits are also mentioned in it. According to Article 9(2) of the regulation, monomers, starting substances, and additives must undergo a risk assessment and receive authorization before being employed in the manufacturing of plastic materials and articles. According to Article 9 of EU Regulation No. 10/2011 mandates that substances utilized in nanoparticle form require authorization, with case-by-case individual assessment by EFSA. Additionally, authorization for conventional substances does not automatically extend to the same substance in nanoform.

Active and intelligent materials play a crucial role in enhancing the shelf life of packaged food and indicating the condition of packaging materials. Regulation (EC) 450/2009 addresses these materials. Intelligent systems are utilized to monitor the state of packaging, and they can be positioned either on the outer surface or separated by a functional barrier, which acts as a barrier within food contact materials. Unauthorized substances can be employed behind this functional barrier, provided they meet migration limits. When considering the application of nanoparticles in active and intelligent materials, a case-by-case assessment is necessary, and they should not be covered under the functional barrier concept.

Some nanoparticles can be added in food contact material like Titanium nitride, Zinc oxide and silver nanoparticles as per EFSA. Details of these nanoparticles are mentioned below:

A. Titanium nitride

In PET bottles titanium nitride can be used up to 20 mg/kg, as no migration of titanium nitride nanoparticle occurs. In PET bottle agglomerates of titanium nitride nanoparticles have a diameter of 100-500nm and primary particles have a diameter of 20nm.

B. Titanium dioxide

Titanium dioxide surface-treated with fluoride-modified alumina can be used at up to 25 % w/w including in the nanoform.

C. Zinc oxide

Zinc oxide nanoparticles, both coated with [3-(methacryloxy)propyl] trimethoxysilane and uncoated, can be used in unplasticized polymers. Specific migration limit [3-(methacryloxy)propyl] trimethoxysilane coated zinc oxide will be 0.05 mg/kg. Zinc oxide can be introduced in nanopowder dispersion in low-density polyethylene plaques at 2%, as substances do not migrate in nanoform, focusing safety evaluation on soluble ionic zinc migration.

D. Montmorillonite clay

Montmorillonite clay modified with hexadecyltrimethylammonium bromide can be used as an additive up to 4.0% w/w in polylactic acid plastics intended for long-term water storage at ambient temperatures or lower.

Montmorillonite clay modified with dimethyldialkyl (C16-C18) ammonium chloride can be used up to 12% w/w in polyolefins in contact with dry foods to which simulant poly(2,6-diphenyl-p-phenylene oxide) is assigned at room temperature or below. The combined specific migration of 1-chlorohexadecane and 1-chlorooctadecane must not exceed 0.05 mg/kg of food.

Both the modified clay forms nano-sized platelets, with one or two dimensions less than 100 nm thick. These platelets should be aligned parallel to the polymer surface and be completely embedded within the polymer.

E. Food contact materials (FCM Nos 998, 859, and 1043)

Substances:

1. **FCM No 998**: (Butadiene, ethyl acrylate, methyl methacrylate, styrene) copolymer, not cross-linked, in nanoform
2. **FCM No 859:** (Butadiene, ethyl acrylate, methyl methacrylate, styrene) copolymer cross-linked with divinylbenzene, in nanoform
3. **FCM No 1043:** (Butadiene, ethyl acrylate, methyl methacrylate, styrene) copolymer cross-linked with 1,3-butanediol dimethacrylate, in nanoform

Usage:

- These substances can be used as particles in non-plasticised PVC up to 10% w/w.
- They are suitable for contact with all food types at room temperature or below, including long-term storage.

Restrictions: When any combination of these substances (FCM Nos 998, 859, and 1043) is used, the total concentration must not exceed upto 10% w/w.

Particle Size:

- The diameter of the particles must be greater than 20 nm.
- At least 95% of the particles must have a diameter greater than 40 nm.

FCM No-1016

FCM No-1016: (methacrylic acid, ethyl acrylate, n-butyl acrylate, methyl methacrylate and butadiene) copolymer in nanoform

Only to be used up to:

a) 10 % w/w in non-plasticised PVC;

b) 15 % w/w in non-plasticised PLA.

The final material shall be used at room temperature or below.

Silver nanoparticles

The EFSA Panel evaluated the safety of silver nanoparticles intended for use in plastics as a surface biocide in contact with various foods. The nanoparticles, ranging from 1 to 100 nm with a mean diameter of 15 nm, remain embedded in the polymer without migration or release under tested conditions, and 99% of particles are below 20 nm. The additive is designed for use as a surface biocide in non-polar plastics, with a recommended concentration of up to 0.025% by weight. It can be applied to various food types and food contact surfaces. The additive does not migrate and resists release through abrasion, ensuring it does not leach into food and raising no toxicological concerns. Soluble silver particles migrated up to 6 μg/kg of food from the surface of food contact materials into the food, well below the group restriction of 50 μg of silver/kg of food proposed by the AFC panel in 2004 (Lambré *et al.*, 2021b).

Canada

Health Canada defines a substance or product as a nanomaterial if it possesses dimensions at or within the nanoscale in one external dimension, or if its internal or surface structure is at the nanoscale. Additionally, it qualifies if it exhibits one or more nanoscale properties or phenomena, regardless of whether it is smaller or larger than the nanoscale in all dimensions (Health Canada, 2011).

Canada has no dedicated regulations for food products related to nanotechnology, regulation related to nanoparticles falls under the purview of existing legislative and regulatory frameworks. Health Canada safeguards and advances health through the utilization of current legislative and regulatory structures to address potential health hazards associated with nanomaterials and to facilitate their health advantages.

Health Canada advises stakeholders to engage early with regulatory authorities, particularly for combination products involving nanomaterials, to evaluate potential risks and benefits. Manufacturers are encouraged to seek pre-submission meetings to discuss necessary safety assessment information regarding the nanomaterials.

Japan

Food nanomaterials are expected to reach 250 billion in the Japanese market by 2030 (Tachikawa, 2012). In March 2010, the Food Safety Commission of Japan (FSCJ) released a report titled "A Fundamental Survey on Safety Evaluation Information Regarding the Utilization of Nanotechnology in Food Industry," offering insights into the implementation of nanotechnology within Japan's food sector (Cho *et al.*, 2010). The document contains a study investigating the potential transfer of synthetic nanoparticles (NPs) from packaging materials to food products through a physicochemical process. Findings indicate that in cases where a polymer matrix with low dynamic viscosity is used and synthetic NPs with a diameter of 2 nm are present in packaging materials, manufactured NPs can migrate into the food. This phenomenon is also applicable to nanocomposites containing silver and polyolefins such as LDPE, HDPE, and PP. In 2013, the METI's Committee on Safety Management for Nanomaterials replaced its previous precautionary measures with a new notice aimed at preventing exposure to nanomaterials (Karim, 2013).

United States of America

The United states government founded the National Nanotechnology Initiative 10 year ago to leverage the research programmes on nanotechnology (Tinkle & Carim, 2011). In October 2011, the National Nanotechnology Initiative published a new strategy aimed at promoting responsible nanotechnology development and aiding regulatory decisions. This strategy, updating the 2008 version, focuses on environmental, health, and safety concerns. It highlights the need for further research into tools to assess the properties of engineered nanomaterials and to detect and monitor their presence in real-world exposure settings (NSET/NEHI, 2011). The FDA is a member agency of the National Nanotechnology Initiative (NNI), a federal research and development program established to coordinate the multi-agency efforts in nanoscale science, engineering, and technology. The goals of the NNI are to maintain a world-class research and development program aimed at realizing the full potential of nanotechnology. The FDA works with international stakeholders, including regulatory agencies, to share perspectives and information on the regulation of nanotechnology products (U.S. Food and Drug Administration, 2013). In June 2011, the FDA released a draft guidance document for public feedback. It aims to assist industries and stakeholders in recognizing situations where the use of nanotechnology might affect regulatory status, safety, effectiveness, or public health impact in FDA-regulated products. The FDA does not keep a record of nanomaterials that it has assessed (FDA, personal communication, 2011).

The research needs identified by the National Nanotechnology Initiative (NNI) for the environmental, health, and safety (EHS) of nanomaterials align with the U.S. Food and Drug Administration (FDA)'s regulatory science research priorities and goals. Below are measurement tools for nanomaterials, in vitro and in vivo assays, and risk characterization, assessment, and communication:

1. Measurement Tools for Nanomaterials:

- NNI Need: Develop tools to measure the physical and chemical properties of engineered nanomaterials (ENMs) in different environments.
- FDA Research Category: Physico-chemical Characterization.
- FDA Goal: Improve methods to detect and measure the properties and safety of nanomaterials in FDA-regulated products, both in biological systems and during manufacturing.

2. In Vitro and In Vivo Assays

- NNI Need: Create lab tests and animal models to predict how nanomaterials will behave in the human body.
- FDA Research Category: Nonclinical Models.
- FDA Goal: Develop and evaluate lab and animal tests to understand the safety and effectiveness of nanomaterials, including how they are absorbed, distributed, metabolized, and excreted in the body.

3. Risk Characterization Information

- NNI Need: Use risk assessment data to understand the hazards, exposure, and risks associated with nanomaterials.
- FDA Research Category: Risk Characterization.
- FDA Goal: Conduct targeted research to improve the understanding of hazards, exposure, and risks in FDA-regulated products containing nanomaterials.

4. Risk Assessment Approaches

- NNI Need: Evaluate methods for assessing and managing risks of nanomaterials.
- FDA Research Category: Risk Assessment.
- FDA Goal: Enhance knowledge and evidence to develop generalized risk assessment approaches for products with nanomaterials.

5. Risk Communication

- NNI Need: Standardize how risk information is communicated within the risk management framework.

- FDA Research Category: Risk Communication.
- FDA Goal: Improve communication about risks associated with FDA-regulated products that contain nanomaterials or are relevant to nanotechnology (U.S. Food and Drug Administration, 2013).

Risk management and assessment: EPA

United States Environmental Protection Agency (EPA) has identified five nanomaterial types for investigation that are widely used in products or have been recognized for their potential uses (EPA, 2011c). The materials being studied are:

1. Carbon tubes and fullerenes
2. Cerium Oxide
3. Titanium dioxide
4. Silver
5. Iron

The approach used for assessing the potential toxicity of nanomaterials.

- Recognize and describe the physical and chemical attributes of engineered nanomaterials.
- Recognize different testing methods and strategies to anticipate human toxicity, including the discovery of biomarkers indicating exposure to nanomaterials or their harmful effects.
- Evaluate the toxicity of nanomaterials in animal subjects, which will involve investigating factors related to host susceptibility and sensitivity that could impact toxicity.

India

In August 2019, the Department of Biotechnology (DBT) under the Union Ministry of Science and Technology in India issued preliminary guidelines aimed at governing the application of nanotechnology within the agricultural domain. These guidelines offer direction on the particular requirements for Nano-based Agricultural Input Products (NAIPs) and Nano-based Food Products (NAPs). While general requirements outlined in the regulations of FSSAI, and BIS apply to all new products, whether nano-enhanced or not, a case-by-case approach is necessary for evaluating NAIPs and NAPs due to the interdisciplinary nature and complexity involved. The guidelines, called Guidelines for Evaluation of Nano-based Agri-input and Food Products in India, aim to secure the quality, safety, and effectiveness of agricultural and food products utilizing nanotechnology. Nanomaterials embedded into various materials like ceramic, cellulose films, and plastic are subject to different kinds of regulation. In the context of evaluating the quality and safety of food products, reference to European Regulation No. (EC) 1935/2004 could provide a framework for assessment. The Great Britain policy 2002/72/EC (14) may be adhered to control plastic and other items that come into contact with food to address concerns about lead and cadmium migration causing food contamination. In 1991, the US EPA provided 12 principles of green chemistry to provide valuable guidance for developing safe nanomaterials in agriculture. These encompass prevention, maximizing atom efficiency, employing less hazardous synthesis methods, creating safer chemicals and solvents, prioritizing energy-efficient design, utilizing renewable resources, minimizing derivatives, promoting catalysis, designing for degradation, conducting real-time pollution analysis, and embracing inherently safer chemistry for accident prevention.

Stability testing of National Agriculture Products

The standard storage stability requirements and procedures for agricultural chemical products can also be utilized for Nano-based Agricultural Input Products (NAIPs) and Nano-based Food Products (NAPs) to guarantee their stability. The storage stability tests outlined in OECD Test Guidelines (TG) 318, FAO/WHO, and APVMA are recommended for adoption. Storage stability tests of materials added with nanoparticles and their test parameters are mentioned below in Table 1 & 2. Safety assessment of manufactured National Agriculture products are mentioned in Table 3.

Table 1: Storage stability tests of National agriculture Products

Sr.No.	Storage stability tests
1	Accelerated storage stability
2	Ambient storage stability
3	Low temperature storage stability
4	Photostability
5	Container compatibility test
6	Stability and transformation dynamics in soil system

Table 2: Tests parameter to analyzed during each storage stability test

Sr.No.	Tests parameter to analyzed during each storage stability test
1	Selection of containers and its testing for transport worthiness
2	Batch
3	Batch and size of products

4	Time period of storage
5	Validation of analytical methods
6	Technical characteristics
7	Microbial stability
8	Reactivity and pH stability

Table 3: Safety assessment of manufactured National Agriculture products

Test Category	Test Methods	Purpose
Dermal exposure/toxicity	ATA	Skin corrosion and irritation
	OECD TG 428	In vitro skin corrosion: Test method for Reconstructed epidermis (RHE)
	OECD TG 431	In vitro skin corrosion: Reconstructed human epidermis (RHE) test method
Eye irritation	OECD TG 437	Bovine corneal opacity and permeability test method for evaluation of chemicals likely to cause eye damage
Inhalation exposure	OECD TG 433	Acute inhalation toxicity for evaluation of health hazard from short-term exposure to a test chemical by inhalation.
	OECD 436	Acute inhalation toxicity: acute toxic class method
Genotoxicity	OECD TG 473	In vitro mammalian chromosomal aberration tes
	OECD TG 476	In vitro mammalian cell gene mutation tests using the hprt and xprt genes
	OECD TG 487	In vitro mammalian cell micronucleus test
Cytotoxicity		ATP Cell Titer-Glo, neutral red uptake, LDH release, MTT, XTT, cell impedance, trypan blue, BrdU, Alamar Blue, WST-1, live/dead cell counting, colony-forming efficiency
	OECD TG 489	In vitro mammalian alkaline comet assay for measuring DNA strand breaks in eukaryotic cells may also be included in the test panel

Environmental safety

1. Aquatic tests (OECD TG 201, 202, 203, 211, 236) are used to assess freshwater organisms' sensitivity to nanoparticles.
2. The U.S. EPA bioassays are used to detect genotoxicity in environmental agents.

3. Soil tests (OECD TG 222, 216, 217, 471) evaluate the chemical effects of nanoparticles on earthworms, soil microbiota, and bacteria.
4. Regulators may also develop and furnish appropriate standard guidelines for assessing the risk to macrobiota.
5. Methods to evaluate the half-life of the NMs in soil may also be included to predict risk.
6. OECD TG 307 may be adopted with necessary adaptations for aerobic-anaerobic transformation of NAPs in soil.
7. OECD section 3; Test 318 In order to assess the environmental risk due to particular NAPs.

Occupational health safety and waste disposal

Exposure management tactics, waste management, and optimal procedures must be followed when dealing with nanomaterials, aligning with the 'Guidelines and Best Practices for Safe Handling of Nanomaterials in Research Laboratories and Industries' formulated by NanoMission, DST, Government of India.

Toxicity evaluations must comply with the principles of Good Laboratory Practices (GLP), ensuring that they are solely carried out by trained and qualified personnel who utilize calibrated and standardized equipment. Strict adherence to written protocols is essential, with Standard Operating Procedures (SOPs) followed diligently for all laboratory and management tasks associated with toxicity evaluation.

Residue analysis

Manufacturers must disclose information regarding nanomaterial residue, the presence of nanosized degradation byproducts in food, or surface coatings applied to food contact materials must be mentioned by the manufacturer during product registration. The report should include details on these aspects.

1. A technique for establishing the detection and quantification thresholds of residues from the applied active ingredient, nanocarrier, and additive.
2. The amounts of produced residues and an overview of the expected risks associated with these residues need to be provided. When it comes to toxicological data, it's important to ensure that there is no transfer of elements from food contact materials, or if there is, that the migrating species are not in nanoparticle form (in which case standard risk assessment procedures must be followed).
3. The bioaccumulation patterns of NAPs need to be assessed by the provided test guideline or an alternative one.

Information required for evaluation of NAPS

The nanofood products are required as per the food safety and standards regulation (FSSAI, 2016). The following information is required for registration of NAPs.

Exposure Risk

Migration of NMs or its degraded products in nan-nanoform from food contact materials should be considered in exposure measurement and hazard characterization are required.

Hazard characterization

Hazard identification and characterization require appropriate in vitro and in vivo studies to determine the fate of NMs. Toxicity testing should be customized as per the exposure.

Migration of food contact materials: When there is no migration from food contact materials, toxicological concerns are negligible. Stringent toxicity studies need to be enforced if there is migration. Both in vitro and in vivo studies need to be done as given by EFSA Scientific Committee, (2011).

Food safety labeling

Every food product that has NMs in it should be labeled as per the Food Safety and Standards (Labeling) regulations, and include information on chemical/ elemental composition of NMs.

Conclusion

The rapid advancement of nanotechnology in food packaging holds immense promise for enhancing food safety, extending shelf life, and reducing waste. However, with these advancements come the critical need for stringent regulatory oversight to ensure the safety of consumers and the environment. Across different regions, regulatory bodies such as the EU's EFSA, Health Canada, Japan's Food Safety Commission, the US FDA and EPA, and India's Department of Biotechnology are actively engaged in evaluating the risks associated with nanomaterials in food contact materials. By implementing comprehensive regulations and guidelines, these authorities aim to address concerns regarding nanoparticle migration, toxicity, and environmental impact. Collaboration between regulators, industry stakeholders, and researchers is vital to establish transparent and effective regulatory frameworks that balance innovation with safety. Ultimately, through careful regulation and oversight, nanotechnology can continue to revolutionize the food packaging industry while ensuring consumer confidence in the safety of their food products.

References

Adholeya, A., & Singh, R. (2020). Guidelines for Evaluation of Nano-Based Agri-Input & Food Products in India. Department of Biotechnology, Ministry of Science and Technology, Government of India.

Anandharamakrishnan, C. (2014). Techniques for formation of nanoemulsions. In Techniques for nanoencapsulation of food ingredients (pp. 7–16). Springer. https://doi.org/10.1007/978-1-4939-1577-1_2

Ashfaq, A., Khursheed, N., Fatima, S., Anjum, Z., & Younis, K. (2022). Application of nanotechnology in food packaging: Pros and Cons. Journal of Agriculture and Food Research, 7, 100270. https://doi.org/10.1016/j.jafr.2022.100270

Barnes CA, Elsaesser A, Arkusz J *et al.* (2008) Reproducible comet assay of amorphous silica nanoparticles detects no genotoxicity. Nano Lett 8:3069–3074. https ://doi.org/10.1021/nl801 661

Bergamaschi E, Bussolati O, Magrini A et al (2006) Nanomaterials and lung toxicity: interactions with airways cells and relevance for occupational health risk assessment. Int J Immunopathol Pharmacol 19:3–10

Chowdhury, P., Gogoi, M., Borchetia, S., & Bandyopadhyay, T. (2017). Nanotechnology applications and intellectual property rights in agriculture. Environmental Chemistry Letters,15(3), 413–419.

Cowburn, R. (2008). Laser surface authentication–reading Nature's own security code. Contemporary Physics, 49(5), 331-342. https://doi.org/10.1080/00107510802341882

DePolo, R., & Baird, J. (2009). Improving polymer composites with magnesium nanosilicate. Polymer Composites, 30(6), 800-807. https://doi.org/10.1002/pc.20554

des Rieux, A., Fievez, V., Garinot, M., Schneider, Y.-J., & Préat, V. (2006). Nanoparticles as potential oral delivery systems of proteins and vaccines: A mechanistic approach. Journal of Controlled Release, 116(1), 1–27. https://doi.org/10.1016/j.jconrel.2006.08.013

Ealia, S. a. M., & Saravanakumar, M. (2017). A review on the classification, characterisation, synthesis of nanoparticles and their application. IOP Conference Series. Materials Science and Engineering, 263, 032019. https://doi.org/10.1088/1757-899x/263/3/032019

European Commission. (2011). Regulation (EU) No 10/2011 of 14 January 2011 on plastic materials and articles intended to come into contact with food. Official Journal of the European Union, L 12, 1-89.

European Parliament and Council. (2004). Regulation (EC) No 1935/2004 of the European Parliament and of the Council of 27 October 2004 on materials and articles intended to come into contact with food and repealing Directives 80/590/EEC and 89/109/EEC. Official Journal of the European Union, L 338, 4-17.

European Parliament, & Council of the European Union. (2009). Regulation (EC) No 450/2009 of the European Parliament and of the Council of 29 May 2009 on active and intelligent materials and articles intended to come into contact with food. Official Journal of the European Union, L 135, 3-11.

FDA (United States Food and Drug Administration). (2011a). Considering whether an FDA-regulated product involves the application of nanotechnology. Draft guidance for industry. Silver Spring, MD, USA: FDA. Retrieved from http://www.fda.gov/RegulatoryInformation/Guidances/ucm257698.htm#rationale

Hasan, S. (2015). A review on nanoparticles: their synthesis and types. Research Journal of Recent Sciences, 1–3. https://www.researchgate.net/publication/273203342

Health Canada. (2011, May 26). Policy Statement on Health Canada's working Definition for nanomaterial. Canada.ca. https://www.canada.ca/en/health-canada/services/science-research/reports-publications/nanomaterial/policy-statement-health-canada-working-definition.html

InsightAce Analytic. (2023). Nanotechnology for food packaging market. Retrieved from https://www.insightaceanalytic.com/report/nanotechnology-for-food-packaging-market/1560.

Kah, M., Beulke, S., Tiede, K., & Hofmann, T. (2013). Nanopesticides: State of knowledge, environmental fate, and exposure modeling. Critical Reviews in Environmental Science and Technology, 43(16), 1823–1867. https://doi.org/10.1080/10643389.2012.671750

Karamfilova, E., Sacher, M., Giulio Sabbati, Ex-Post Impact Assessment Unit, & Publications Management and Editorial Unit. (2016). Food Contact Materials - Regulation (EC) no 1935/2004. In European Implementation Assessment (Report PE 581.411). European Parliament. https://doi.org/10.2861/162638

Karim, M. E., Munir, A. B., Reza, A. W., Muhammad-Sukki, F., Yasin, S. H. M., Abu-Bakar, S. H., *et al.* (2015). Too enthusiastic to care for safety: Present status and recent developments of nanosafety in ASEAN countries. Technological Forecasting and Social Change, 92, 168–181.

Khan MI, Mohammad A, Patil G et al (2012) Induction of ROS, mitochondrial damage and autophagy in lung epithelial cancer cells by iron oxide nanoparticles. Biomaterials 33:1477–1488. https ://doi.org/10.1016/j.bioma teria ls.2011.10.080

Lambré, C., Baviera, J. M. B., Bolognesi, C., Chesson, A., Cocconcelli, P. S., Crebelli, R., Gott, D. M., Grob, K., Lampi, E., Mengelers, M., Mortensen, A., Steffensen, I., Tlustos, C., Van Loveren, H., Vernis, L., Zorn, H., Castle, L., Di Consiglio, E., Franz, R., . . . Rivière, G. (2021b). Safety assessment of the substance silver nanoparticles for use in food contact materials. EFSA Journal, 19(8). https://doi.org/10.2903/j.efsa.2021.679

Luo C, Urgard E, Vooder T, Metspalu A (2011) The role of COX-2 and Nrf2/ARE in anti-inflammation and antioxidative stress: aging and anti-aging. Med Hypotheses 77:174–178. https ://doi. org/10.1016/j.mehy.2011.04.002

Mohammadi, Z., & Jafari, S. M. (2021). Regulatory principles on food nanoparticles legislated by Asian and Oceanian countries. In Elsevier eBooks (pp. 201–238). https://doi.org/10.1016/b978-0-12-815725-1.00006-9

Mulvaney, P. (2015). Nanoscience vs Nanotechnology—Defining the Field. ACS Nano, 9(3), 2215–2217. https://doi.org/10.1021/acsnano.5b01418

NSET/NEHI. (2011). NNI 2011 environmental, health, and safety (EHS) research strategy. United States National Nanotechnology Initiative, Nanotechnology Environmental and Health Implications Working Group of the Nanoscale Science, Engineering, and Technology Subcommittee. Retrieved from http://www.nano.gov/node/681

OECD. (2009). Preliminary review of OECD test guidelines for their applicability to manufactured nanomaterials. Paris. ENV/JM/MONO(2009)21. Retrieved from http://www.oecd.org/officialdocuments/publicdisplaydocumentpdf/?doclanguage=en&cote=env/jm/mono(2009)21

OECD. (2014). Report of the OECD expert meeting on the physical chemical properties of manufactured nanomaterials and test guidelines. ENV/JM/MONO(2014)15. Retrieved from http://www.oecd.org/officialdocuments/publicdisplaydocumentpdf/?cote=env/jm/mono(2014)15&doclanguage=en

OECD. (2015). Guidance manual towards the integration of risk assessment into life cycle assessment of nano-enabled applications series on the safety of manufactured nanomaterials No. 57. ENV/JM/MONO(2015)30. Retrieved from http://www.oecd.org/officialdocuments/publicdisplaydocumentpdf/?cote=env/jm/mono(2015)30&doclanguage=en

OECD. (2017). Test No. 318: Dispersion Stability of Nanomaterials in Simulated Environmental Media. OECD Guidelines for the Testing of Chemicals, Section 3. OECD Publishing, Paris. https://doi.org/10.1787/9789264284142-en

Pan, K., & Zhong, Q. (2016). Organic nanoparticles in foods: fabrication, characterization, and utilization. Annual Review of Food Science and Technology, 7(1), 245–266. https://doi.org/10.1146/annurev-food-041715-033215

Peters, R., Brandhoff, P., Weigel, S., Marvin, H., Bouwmeester, H., Aschberger, K., *et al.* (2014). Inventory of nanotechnology applications in the agricultural, feed and food sector. EFSA Supporting Publications, 11(7), 621E. https://doi.org/10.2903/sp.efsa.2014.EN-621

Regulation (EC) No 1935/2004 of the European Parliament and of the Council of 27 October 2004 on materials and articles intended to come into contact with food and repealing Directives 80/590/EEC and 89/109/EEC. Retrived from: https://eurlex.europa.eu/LexUriServ/LexUriServ.do?uri=OJ:L:2004:338:0004:0017:en:PDF

Tachikawa, M. (2012). Uncertainty of, and stakeholder response to, emerging technologies: food nanotechnology in Japan. Ethics in Science and Environmental Politics, 12(2), 113–122

Tinkle, S. & Carim, A. 2011. Responsible development of nanotechnology: maximizing results while minimizing risk. Washington, DC: Office of Science and Technology Policy. Retrieved from (http://www.whitehouse.gov/blog/2011/10/20/responsible

U.S. Food and Drug Administration. (2013). Nanotechnology Regulatory Science Research Plan. Retrieved from https://www.fda.gov/media/87265/download

World Health Organization & Food and Agriculture Organization of the United Nations. (2010). Manual on the development and use of FAO and WHO specifications for pesticides (2nd rev. of the 1st ed.). Rome. Retrieved from http://www.fao.org/3/a-y4353e.pdf

10

Safe Inks: The Missing Piece in Sustainable Packaging

Jatin Takkar

Head, Product Safety & Regulatory, Siegwerk, India

Packaging is an essential tool to make products stand out from the competition. Beyond the shelf appeal, packaging also provides protection to the contents from physical, chemical and microbiological hazards. However, a lesser-known fact is that the same packaging material could also be a source of chemical contaminants, and hence, may impact food safety as well as food quality. Among the different components of packaging material, packaging inks play a critical role as they might hamper product and consumer safety.

Each packaging ink is a mixture of different chemicals and some of these chemicals might be harmful with the potential to migrate to the foodstuff. In terms of risk, the chemicals could be toxic, carcinogenic, mutagenic, repro-toxic and even endocrine disruptors. The chemical contaminants from the packaging inks can get transferred to the foodstuff by a process called migration.

Over the last two decades, a number of scientific papers have been published, revealing the migration potential of substances from the printed matter. However, the first big scandal that brought the subject alive was about the findings of isopropyl thioxanthone (ITX, a low molecular photo initiator used in UV inks) in baby milk and other liquid foodstuffs all over Europe, resulting in a product recall in the year 2005. The food scandal alerted the packaging chain about the linkage between packaging inks and food safety. Consequently, the authorities issued strict regulations for the application of printing inks as well as general food control measures.

Food Safety and Migration

Migration can be described as a transfer of chemical contaminants from the food packaging material to the food, thereby making it unsafe. Migrants are substances which, due to their chemical characteristics and molecular size, move from a printed layer into the packed food. There are different types of migration, which are as follows:

- **Set-off migration**

Migrations can happen from one layer to another, such as a surface printed layer to the non-printed food-contact surface, which is later brought into contact with food. If these are in direct or close contact, like in a reel or a stack after printing, set-off migration can easily occur due to the pressure existing in the reel or stack.

- **Diffusion migration**

Small and mobile molecules can easily penetrate and diffuse across packaging material layers. This can occur even if the printed material has not yet been converted into a food package and filled with food, or later when the printed package is filled with food and the food starts to extract the migrants from the packaging material.

- **Gas-phase migration**

Migration can also happen from a cardboard (the releasing reservoir) via the gas phase within the pack, to end up in food which acts as the recipient reservoir (gas phase migration). This can, for example, occur with migrants like mineral oils or some UV photo-initiators that might not be generally known as being volatile, such as organic solvents.

This process of migration is influenced by several factors like the physico-chemical properties of the migrant, of packaging material and of the food product; the storage time and temperature, and the size of packaging in proportion to foodstuff volume.

Concept of safe, compliant and migration optimized inks

Compliance in packaging safety needs to be based on these three pillars — legal requirements; brand owner requirements and self-commitments.

Each supply chain partner (including the ink manufacturer, adhesive manufacturer and substrate manufacturer) needs to be aware of the legal requirements (relevant for a particular region) for their component of the packaging structure and must build systematic processes to ensure delivery of consistent and safe products. If the packaging material needs to be exported, the legal requirements of the importing country must also be given due consideration.

Brand owners, with their global footprints, sometimes have their own specific requirements for packaging safety beyond the legal requirements of the countries (especially developing countries like India). Each supply chain partner needs to make themselves aware of such requirements and upgrade their products to comply with such norms, in order to offer the same level of consumer safety across all regions.

Each supply chain partner is responsible for their own products and formulations. Hence, it becomes the ethical responsibility of supply chain partners to deliver safe products (with only safe chemicals in their formulations) and also update the authorities and brand owners on such initiatives to benefit the society at large.

Compliance can never be achieved through an isolated approach since each packaging structure is a mixture of different components (including substrate, ink and varnishes), and hence, there should be an open and transparent communication across the supply chain on the requirements, as well as on the risk assessment of the individual components of the packaging structure.

Need for Migration-optimized Inks

In the Indian market, there is a demand for the so-called non-toluene (NT) inks, non-toluene non-ketone (NTNK) inks, mineral oil-free inks, etc. However, safe inks should not be misled with such terminology adopted by the market, as such inks may not be safe.

Each printing ink consists of four general components-binder/resin, solvent, additives and pigment/colorant. By excluding a solvent like toluene, ketone, mineral oils, etc, one cannot assume and be assured that the inks are safe.

Safe inks (or migration-optimised inks) are inks with careful selection of raw materials in all four categories, so that none of the components of the printing ink formulations impact consumer safety.

Hence, the next time one is looking for safe inks, he/she should ask for migration-optimised inks.

Good Manufacturing Practices

Packaging materials and articles shall be manufactured in compliance with good manufacturing practices (GMPs), so that, under normal or foreseeable conditions of use, they do not transfer their constituents to food in quantities which could endanger human health; bring about an unacceptable change in the composition of the food or bring about a deterioration in the organoleptic characteristics. GMPs should be exercised across the supply chain if packaging safety needs to be ensured.

There are several parameters and conditions that need to be controlled at the converters' shop floor to make sure that the final intended article is safe. Even if one has procured the safe raw materials (such as substrate, inks, adhesives etc.), if they are not processed in the right manner, the final intended article would never be food safe.

For example, in the case of solvent-based inks, solvent drying is an essential part of the manufacturing process of the final packaging material, which is carried out at the converter's end. In case solvent drying is not handled appropriately at the converter shop floor, retained solvent in the final packaging material can migrate into the product, thereby not only endangering human health, but also impacting the organoleptic properties of the product.

Converters also need to ensure that cross-contamination at the facility is well taken care of and there are adequate validated controls to ensure that.

Changing Regulatory Landscape

In India, the Food Safety and Standards Authority of India (FSSAI) is responsible for protecting and promoting public health through the regulation and supervision of food safety. FSSAI has been established under the Food Safety and Standards Act, 2006. Food Safety and Standards Authority of India (FSSAI) has raised the bar on packaging safety by updating the norms on packaging materials as per Food Safety and Standard (Packaging) Regulations, 2018 and its subsequent ammendments. FSSAI has given the primary responsibility to food business operators to ensure that the packaging material complies with the Food Safety and Standard (Packaging) Regulations, 2018. FSSAI also defined the term " Food Grade". The standard clearly indicates that a packaging material is to be considered food grade when the material is made of substances which are safe and suitable for their intended use, and shall not endanger human health or result in unacceptable change in the composition of the food or organoleptic characteristics.

Packaging regulation prescribed by FSSAI is a more stringent version of the previous regulation and now encompasses more elements such as paper, metal, packaging inks etc.FSSAI now prescribes that printing inks for use on food packages shall conform to IS 15495. FSSAI also has prohibits the use of newspaper or any such material used for storing and wrapping of food. Additionally , FSSAI has also prohibited the direct contact between the printed surface of packaging material and the food product.

Packaging Inks conformance to IS 15495

The Indian Standard IS 15495:2004 'Printing Ink for food packaging-Code of Practice' prescribes guidelines for printing inks for use in food packages. The standard differentiates between four categories of printing inks and gives guidance on the formulation of the respective inks:

1. **Printing inks on external (secondary/tertiary) food packaging**: They can be formulated freely, but must not contain substances from the exclusion list and must not contain toxic substances. In the case, that a functional barrier does not exist, bleeding dyes and coloring agents need to be avoided.
2. **Printing inks on immediate food wrappings** : Those must be applied to the outside of the food wrapper, comply with the exclusion list and must not contain toxic substances. Inks are to be printed in such a manner as to avoid set-off. The final intended articles need to be manufactured in a manner that under normal or foreseeable condition of use, they shall not transfer their constituents to the food in quantities, which may endanger human health, cause a deterioration in the organoleptic characteristics or an unacceptable change in the nature, substance and/or quality of the food. In the case, that a functional barrier does not exist, bleeding dyes and coloring agents need to be avoided.
3. **Printing inks for direct food contact** : They must be formulated only with food additives under the appropriate regulation of the Government of India. The final intended articles need to be manufactured in a manner that under normal or foreseeable condition of use, they shall not transfer their constituents to the food in quantities, which may endanger human health, cause a deterioration in the organoleptic characteristics or an unacceptable change in the nature, substance and/or quality of the food. In the case, that a functional barrier does not exist, bleeding dyes and coloring agents need to be avoided.
4. **Printing inks for disposables (e.g. paper plates, drinking straws or table napkins):** Those inks must not contain substances from the exclusion list or those, which are otherwise known to be toxic. Moreover, they shall be formulated to avoid bleeding onto the food.

Specified Requirements Across the Supply Chain Stakeholders

a) Ink Manufacturer

- Ink manufacturers are responsible to formulate packaging inks by using only raw materials other than those known to be toxic, carcinogenic, sensitizing or mutagenic, primarily governed by the exclusion list as per Annex A.
- Ink manufacturers are expected to take all necessary precautions to meet the guidelines of the standard.
- Ink manufacturers shall inform the converter and print buyers on the suitability of ink type towards the packages of food and the norms

followed in formulations as per the requirement.

b) Printer / Converter:

- The printer and converter is finally responsible for manufacturing and storage of the food packages in such a manner by which all preventable transfer of material from the ink or coating to the food content is avoided, even if such transfer is unobjectionable on the grounds of health, odor and flavor.
- The printer needs to ensure that the storage environment should be free from potential volatile contaminants, which could adversely affect the organoleptic characteristics of the food.
- For immediate food wrappings and direct contact applications the final intended articles needs to be manufactured in such a way, that under normal or foreseeable condition of use, they shall not transfer their constituents to the food in quantities, which may endanger the human health, cause a deterioration in the organoleptic characteristics or an unacceptable change in the nature, substance and/or quality of the food.
- For immediate food wrappings and direct contact applications the printer needs to establish appropriate controls to avoid set-off in the printing process.
- The converter needs to perform adequate analysis for the specific application in context of validating the bleeding capacity of dyes and pigments used in the inks.

c) Print Buyer/Brand Owner:

- Print Buyers need to design the packaging with the restriction of printing in mind.
- For immediate food wrappings and direct contact applications the final intended articles needs to be manufactured in such a way, that under normal or foreseeable condition of use, they shall not transfer their constituents to the food in quantities, which may endanger the human health, cause a deterioration in the organoleptic characteristics or an unacceptable change in the nature, substance and/or quality of the food.
- Relationship between the press speed and the curing/drying power needs to be fully understood to ensure that an adequate curing/drying takes place.
- For immediate wrappers, the print buyer needs to recommend substrates with sufficiently low permeability to prevent migration.

Balancing safety and sustainability is crucial in today's packaging landscape. While adhering to, and even advocating for, stricter regulations is essential, fostering collaboration and transparency throughout the supply chain holds equal weight.

The rise of the Circular Economy further elevates the importance of packaging safety. As brands pursue eco-friendly initiatives – whether it's using thinner plastic layers, switching materials, adopting mono-material packaging, or incorporating recycled content – the potential for increased chemical migration grows. This underscores the absolute need for due diligence in safeguarding packaging material safety.

The key takeaway? Sustainable practices and safety must work in tandem. Only when these two pillars are equally prioritized can we ensure a positive impact on both people and the planet.